Veterinary Andrology and Artificial Insemination

NIPA® GENX ELECTRONIC RESOURCES & SOLUTIONS P. LTD.
New Delhi-110 034

Veterinary Andrology and Artificial Insemination
Concepts and Applications

Vikas Sachan
Assistant Professor, Department of Veterinary Gynecology and Obstetrics
College of Veterinary Sciences and Animal Husbandry
UP Pandit Deen Dayal Upadhyaya Pashu Chikitsa Vigyan Vishwavidyalaya
Evam Go Anusandhan Sansthan (DUVASU)
Mathura - 281001, Uttar Pradesh, India

NIPA® GENX ELECTRONIC RESOURCES & SOLUTIONS P. LTD.
New Delhi-110 034

NIPA® GENX ELECTRONIC
RESOURCES & SOLUTIONS P. LTD.

101,103, Vikas Surya Plaza, CU Block
L.S.C.Market, Pitam Pura, New Delhi-110 034
Ph. +91 11 27341616, 27341717, 27341718
E-mail: newindiapublishingagency@gmail.com
www: www.nipabooks.com

For customer assistance, please contact
Phone: + 91-11-27 34 17 17
Fax: + 91-11-27 34 16 16
E-Mail: feedbacks@nipabooks.com

ISBN: 978-81-19235-31-5

Composed and Designed by NIPA®.

उ.प्र. पं. दीनदयाल उपाध्याय पशु चिकित्सा विज्ञान विश्वविद्यालय
एवं गो अनुसंधान संस्थान, मथुरा-281001 (उ.प्र.)

U.P. Pandit Deen Dayal Upadhyaya Pashu Chikitsa Vigyan Vishwavidyalaya
Evam Go Anusandhan Sansthan (DUVASU), Mathura-281001 (U.P.)

प्रो.(डॉ) ऐ.के. श्रीवास्तव
कुलपति
Prof. (Dr. A.K. Srivastava)
Vice-Chancellor

Foreword

The valuable teaching and Scientific resource is not only required for researchers and academicians but also for practitioners and students who aspire to contribute to the enhancement of animal health and productivity. In today's rapidly evolving world, where the demand for high-quality livestock production is paramount, the intersection of veterinary andrology and artificial insemination has become an indispensable area of study. This book is meticulously curated and authored by esteemed experts, delves deep into the intricacies of animal reproduction. Here attempt has been made for exploring the fundamental principles and cutting-edge techniques that underpin successful artificial insemination programs. As such this book will be very useful for students, teachers and Veterinary Practicioners. 1 am very happy to note that one of my distinguished faculty Dr. Vikas Sachan has brought this book to address the seamless transitions into the realm of male reproduction and artificial insemination, presenting a comprehensive overview of the different tools, and technologies employed to ensure effective breeding programs. This book bridges the gap between theoretical knowledge and practical application of the male reproductive system and the insemination in farm and companion animals with a holistic approach. The authors provide invaluable insights that will undoubtedly empower veterinarians, animal scientists, and agricultural professionals to optimize their breeding practices while keeping sustainability and ethical considerations at the forefront. I would like to express my heartfelt appreciation to the author and contributors for their dedication and expertise in compiling this invaluable resource. I also feel that this book will be supportive for veterinary graduates and postgraduates to ensure a quick conceptualization of the fundamentals related to male reproduction and AI as per their syllabus. In conclusion, this book "Veterinary Andrology and Artificial Insemination: Concepts and Applications" stands as a testament to the relentless pursuit of knowledge and excellence in the field of animal reproduction.

(A.K. Srivastava)

Tel.: 0565-2470199; Fax : 0565-2470819; Mob. 9466592661; E-mail: duvasuvc@gmail.com; https://www.upvetuniv.edu.in

Preface

Infertility, not only in farm animals but also in companion animals has remained a burning issue for many decades. Although female infertility is often considered the primary culprit, it is crucial to recognize that male animals are equally responsible for this issue. One can say that the reproductive efficiency of the male animal plays a decisive role to get the maximum productivity from female animals. In the same context, it is very important to study all the aspects related to the bio-physiology of the male reproductive system to understand the causes of male infertility. By comprehending the intricacies of male reproductive biology, andrologists, and clinicians can develop effective remedies to address the issue of infertility in male animals. Moreover, addressing the problem of male infertility in animals along with better use of artificial insemination can have far-reaching benefits, such as improving the production quality of livestock and increasing the availability of companion animals for adoption.

This book aims to present a comprehensive guide on the male reproductive system, male infertility, semen evaluation, cryopreservation, and artificial insemination in animals. It provides a concise and interactive sequence of information related to veterinary andrology and artificial insemination in a reader-friendly format. It is tried to write the chapters in very simple language and presented in bullet points to keep the reader engaged with interest. The inclusion of various figures and arrow diagrams at appropriate places in the text makes it easier for the reader to understand the complex concepts in a better way. This book has compiled the related facts which will help undergraduate and postgraduate students in building robust concepts and fundamentals of veterinary andrology and artificial insemination. This book will also help the aspirants to be prepared which are going to appear for various examinations of veterinary science for higher studies. By providing a clear and concise guide to veterinary andrology and artificial insemination, this book aims to contribute to the advancement of animal welfare and agricultural productivity.

Vikas Sachan

Preface

Acknowledgement

The author extends his heartfelt gratitude to all brains who directly or indirectly gave their valuable contributions and whose invaluable contributions have shaped this comprehensive work on presenting book. Their expertise and unwavering support have played a pivotal role in the creation of this work.

Professor (Dr.) P K Shukla, your profound knowledge and passion for veterinary field have illuminated the pages of this book. Your insightful inputs, directions, and scholarly guidance have been instrumental in ensuring the accuracy and depth of the content.

I would like to express my sincere gratitude and thanks to Professor (Dr.) Atul Saxena, whose experience in the field of animal reproduction undoubtedly elevated the quality of this publication. I admit my extended thanks for your role as a pillar of support with amicable and positive directions which help to give shape to this book. I express my genuine appreciation to Dr. Anuj Kumar alongwith all my mentors, senior and junior colleagues for their relentless support, inspiration and moral support. I am also thankful to Drs. Sanjay Mishra and post graduate students Drs. Avaneesh, Arun Kumar, Mohit chaudhary, Garima Singh, Neha Chaudhary for their great support and help. Thank you for your insightful comments and erudite suggestions which make the work possible.

Words cannot describe the encouragement and inculcations given by my parents which help me at every stage in writing this book. I would like to acknowledge, my wife Dr. Roshni Sachan and lovely son Keyansh Sachan for constant encouragement sacrifices, untiring patience which helped me most to complete the task.

To the readers, I present this book with the knowledge that it has been crafted with utmost care and expertise, thanks to the invaluable inputs from all my mentors. I hope that the knowledge shared within these pages proves to be enlightening and beneficial, advancing the understanding of Veterinary Andrology and Artificial Insemination in the academic and professional spheres.

Author

Contents

List of Contributors

Abhinov Verma, Assistant Professor (Department of Veterinary Anatomy), U.P. Pandit Deen Dayal Upadhyaya Pashu-Chikitsa Vigyan Vishwavidyalaya Evam Go Anusandhan Sansthan (DUVASU), Mathura-281001 U.P. India

Akhil Patel, Assistant Professor (Department of Veterinary Gynaecology and Obstetrics), Sardar Vallabhbhai Patel University of Agriculture & Technology, Meerut-250110 (U.P.)

Ajit Singh Mahla, Scientist (Physiology and Biochemistry Division), ICAR-CSWRI, AvikaNagar, Tonk,Rajasthan, India

Anuj Kumar, Associate Professor (Department of Veterinary Gynaecology and Obstetrics), U.P. Pandit Deen Dayal Upadhyaya Pashu-Chikitsa Vigyan Vishwavidyalaya Evam Go Anusandhan Sansthan (DUVASU), Mathura-281001 U.P. India

Atul Saxena, Professor (Department of Veterinary Gynaecology and Obstetrics), U.P. Pandit Deen Dayal Upadhyaya Pashu-Chikitsa Vigyan Vishwavidyalaya Evam Go Anusandhan Sansthan (DUVASU), Mathura-281001 U.P. India

Avneesh Singh, MVSc. (Department of Veterinary Gynaecology and Obstetrics), U.P. Pandit Deen Dayal Upadhyaya Pashu-Chikitsa Vigyan Vishwavidyalaya Evam Go Anusandhan Sansthan (DUVASU), Mathura-281001 U.P. India

Chetna Ganwar, Scientist (Animal Reproduction), CIRG, Makhdoom, Farah, Mathura, U.P. India

Dileep Kumar Yadav, Assistant Professor (Department of Veterinary Clinical Complex), RPS, COVSc., Balana, Mahendragarh, Haryana, India

Jitendra Kumar Agrawal, Assistant Professor (Department of Veterinary Gynaecology and Obstetrics), U.P. Pandit Deen Dayal Upadhyaya Pashu-Chikitsa Vigyan Vishwavidyalaya Evam Go Anusandhan Sansthan (DUVASU), Mathura-281001 U.P. India

Hanuman Prasad Yadav, Assistant Professor (Department of Veterinary Gynaecology and Obstetrics), U.P. Pandit Deen Dayal Upadhaya Pshu-Chikitsa Vigyan Vishwavidyalaya Evam Go Anusdhan Sansthan (DUVASU), Mathura - 281001, U.P., India

Priya Ranjan Kumar, Assistant Professor (Department of Veterinary Gynaecology and Obstetrics) Faculty of Veterinary and Animal Sciences, Institute of Agricultural Sciences RGSC, BHU, Barkachha, Mirzapur-231001(U.P.), India

Satish Pathak, Assistant Professor (Department of Veterinary Anatomy), Faculty of Veterinary and Animal Sciences, Institute of Agricultural Sciences, RGSC, BHU, Barkachha, Mirzapur-231001(U.P.)

List of Contributors

1

Functional Anatomy of the Male Reproductive System

Abhinov Verma and Satish Pathak

Anatomically the male reproduction system comprises of testes, scrotum, duct system, accessory sex glands, prepuce, and penis. Production of spermatozoa and male sex hormone are the main functions of male reproductive physiology. The Male reproductive system can be categorized as:

- Primary sex organ – Testicles
- Secondary sex organ – Epididymis, vas deference, urethra, and penis.
- Accessory sex glands – Seminal vesicles, prostate, and bulbourethral glands/cowper's gland.

1. Testes (Paired male gonads) (Singular: Testis or Testicle)

- Testes are the primary sex organs. They are situated in the scrotum at the inguinal region with the help of tunics and spermatic cord. Opening of the scrotal pouch into the abdominal cavity is called as inguinal canal. The testis is covered with an extra abdominal tunic named as tunica vaginalis, tunica albuginea, and tunica vasculosa from outside to inward. The fluid filled in between the parietal and visceral layers of tunica vaginalis is responsible for the movement of the testis within the scrotum. The upper end of a testicle is occupied by the caput epididymis and the lower end is connected to the cauda epididymis.
- **Comparative parameters of testis in different species**

Sr.No.	Species	Plane of testis	Shape of testis	Testicular diameter	Weight of testis (gm)
1	Cattle	Vertical	Elongates oval	5-9 cm	200-550
2	Ram	Vertical	Elongates oval	4 cm	200-400
3	Buck	Vertical	Elongates oval	4 cm	200-400

Sr.No.	Species	Plane of testis	Shape of testis	Testicular diameter	Weight of testis (gm)
4	Stallion	Horizontal	Oval	4-8 cm	150-300
5	Boar	Oblique	Elliptical	5-9 cm	500-800
6	Dog	Oblique	Round to oval	1-3 cm	8-15
7	Tom	Oblique	Round to oval	0.7–1.5 cm	1-2

- In buffalo bull, the testis and scrotum are smaller, and the penile sheath is less pendulous than in cattle bull. Body weight is positively correlated with testicular weight, testicular volume, total epididymal mass, and total scrotal width in normal male animals especially in dog.
- Endocrine function of the testis is the production of testosterone by leydig cells under the influence of LH/ICSH. The exocrine function of the testis is spermatogenesis under the influence of FSH & testosterone. Functions of all accessory sex glands are under the control of androgen.
- Castration (Removal of testicles) – It reduces the aggressive behaviour of a male animal. In boar, it reduces the boar taint and improves the carcass quality.
- **Cremaster muscle** (striated) attaches from internal abdominal oblique muscle to vaginal tunics. Lowering down and raising up of testes in response to environmental temperature or any abnormal stimulus is the key function of these muscles to maintain the testicular temperature. Cremaster muscles are responsible for retracting the testes within the abdominal cavity of the elephant, deer, rabbit, etc. during their non-breeding season.
- Testes are held in the scrotum with the help of the **spermatic cord** which passes through the inguinal canal. The spermatic cord comprises –
    - Internal spermatic artery
    - Spermatic vein (pampiniform plexus)
    - Lymphatic vessels
    - Autonomic nerves from renal and mesenteric plexus
    - Internal cremaster muscle
    - Vas deferens
    - Tunica vaginalis propria
- All components of the spermatic cord are enclosed within vaginalis tunics.

- **Mesorchium** is the layer of peritoneum connecting the visceral and parietal layers of vaginal process that covers the blood and lymphatic vessels, nerves that supply the testicle, and the epididymis. Peritoneal fold that holds the epididymis is called the **mesoepididymis**.

2. Scrotum

- It is the cutaneous pouch that encloses the testes. The Testes are fixed in scrotum with the help of the **scrotal ligament** (attached with vaginal tunics and scrotal fascia). Scrotal skin bears a small number of hairs except in buck, ram, stallion, and tom.
- There are several layers in the scrotum over testes – (outward to inward) Epidermis, dermis, **tunica dartos**, tunica vaginalis communis or processus vaginalis or tunica vaginalis reflexa (Extension of parietal peritoneum), tunica vaginalis propria (Extension of visceral peritoneum), **tunica albuginea** (fibrous layer with some smooth muscles), tunica vasculosa (layer with blood vessels).
- Tunica dartos is the most prominent layer and has fibroelastic connective tissues and smooth muscles. Tunica dartos forms an intertesticular septum to divide the scrotum into two halves. Tunica vaginalis is an extension of the peritoneum passing through the abdominal wall at inguinal canal. Space between tunica vaginalis communis and tunica vaginalis propria, there is a cavity of the vaginal process. Tunica vaginalis is closely opposed to the tunica albuginea of the testis but separated by a potential space termed the vaginal cavity. Tunica vasculosa has blood vessels that spread and supply blood vasculature all over the testis.
- Scrotum also participates in thermoregulation of the testes (for spermatogenesis). In stallion and dog, the scrotal skin contains large numbers of sebaceous and sweat glands, which assist in thermoregulation of the testicles.
- In all species, they are situated between the thighs except in boar and tom where they are positioned caudal to thighs.
- The scrotum of the horse is relatively short and non-pendulous when compared with that of other farm animals such as bull or ram. Testes in camel are situated in non pendulous scrotum located in the perineal region.

The duct system from inside to outside of the testicle is: Seminiferous tubules – Rete testes – Efferent duct (13-20 in nos.) – Caput epididymis – Corpus epididymis – Cauda epididymis – Vas deferens – Ampulla – Pelvic urethra – Penile urethra – Penile exit

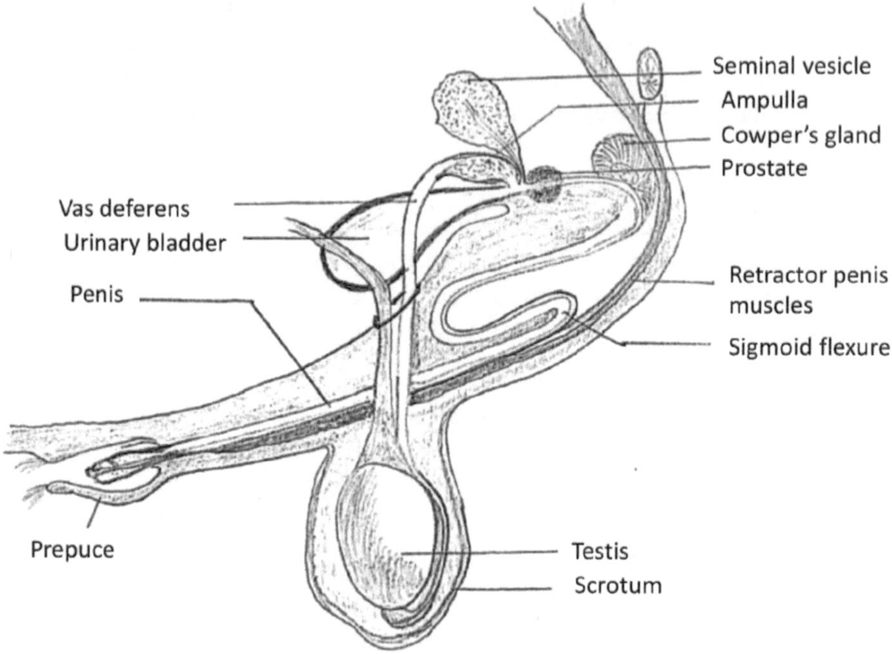

Bull reproductive system

3. Seminiferous Tubule

- Extension of tunica albuginea progresses and form mediastinum at the centre of testes with dividing testicular parenchyma into lobules (encloses seminiferous tubules) and joins at mediastinum. Seminiferous tubules are highly convoluted, unbranched, and open at both ends. They are about 75% of the total testicular mass.
- Seminiferous tubules have somatic Sertoli cells (sustentacular / nurse cells) and sperm-producing germ cells (spermatogonial germ cell line).
- Sertoli cells are named after Enrico Sertoli and Leydig cells are named after its discoverer Franz Leydig.
- **Sertoli cells** secrete mullerian inhibiting factor/hormone (MIH) which inhibits the development and differentiation of the paramesonephric or mullerian duct system. The number of sertoli cells doesn't increase after puberty i.e., it limits spermiogenesis or spermatogenic yield. Sertoli cells form the blood testes barrier just before puberty to separate differentiating germ cells from blood circulation.

- Testosterone-producing **leydig cells** (interstitial cells) along with blood vessels and lymphatics are present in interstitial space i.e., in between the seminiferous tubules. In ram leydig cells are found in clusters around blood vessels.
- Length of seminiferous tubules in testes:

  Boar – 6 Km

  Bull – 5 Km

  Ram – 4 Km

  Dog – 150 m

  Tom – 25 m

4. Rete Testes

- Seminiferous tubules open into collecting tubules which open into rete testes. Rete testes lined by non-secretory cuboidal epithelial cells. Rete testis is absent in dog so collecting tubules open directly into vas efferens. In ungulates and carnivores, the rete testis is located at the center of the testis within the fibrous matrix of mediastinum.

5. Efferent Ductules (Vas efference)

- Vas efferens or ductus efferens connect rete testes to the epididymis. These are 13 to 20 in number in bull and 13 to 15 in the stallion. Spermatophagy occurs in the efferent ductules and possibly the epididymis.

6. Epididymis

- It is a highly convoluted tube that is firmer than testes in consistency. Epididymis can be divided into the head (caput), body (corpus), and tail (cauda). It is lined by a circular muscle fiber layer and a pseudostratified columnar cell layer. Histologically epididymis can be divided into three parts (not related to caput, corpus, and cauda). The initial segment has ciliated cells with large kinocilia and stereocilia. The length of cilias becomes shortened toward the terminal part along with lumen becomes widened. **Medusa cells** seen in the semen are the ciliated cells from caput epididymis.
- Length of the epididymis –
  - Bull – 30-36 m
  - Ram – 50 m

  - Boar – 50-54 m
  - Stallion – 20 m

- In stallion and dog, the caput epididymis is situated cranially, corpus epididymis dorsally, and cauda epididymis caudally to the testicle while in ruminants caput, corpus, and cauda epididymis is situated dorsally, caudally and ventrally to testicle respectively.
- Functions of the epididymis are regulated by androgen –

  a. **Absorption** of excess fluid occurs in caput and cauda epididymis. It increases sperm concentration. More than 90% of fluid leaving the testis is absorbed in the efferent ducts and head of epididymis.

  b. **Secretions** tend to maintain sperm viability during storage. Maximum protein secretion occurs in initial segment of the epididymis which is required for sperm flagellar activity, metabolic process, motility, and fertilization capacity. It secretes sialic acid, carnitine, GPC (glycerylphosphorylcholine), Immobilin and Quescent factor (in bull). There is a high level of 5α reductase (Convert testosterone into dihydrotestosterone DHT) in epididymis, especially in the initial segment.

  c. **Storage** occurs mainly in cauda epididymis. It stores matured sperms (with extended viability)

     - 50% of extragonadal sperm (66% in stallion)
     - 75% of total epididymal sperms
     - Nearly 75000 million sperms

  d. **Maturation** generally occurs in caput and corpus epididymis. Proximal cytoplasmic droplet converts into the distal cytoplasmic droplet during sperm maturation. In the epididymis, there is a modification in surface glycoproteins of the sperm plasma membrane to reduce surface immunogenicity which enhances the ability to bind with zona pellucida. After maturation, spermatozoa are maintained in a protected, quiescent state because of the luminal environment and cool temperature in the cauda epididymis. Here, spermatozoa acquire motility and fertilizing capacity. The fertility of sperms in cauda epididymis is more than those in corpus epididymis.

e. **Transportation** of spermatozoa is carried out by cilia movement and peristalsis of smooth muscle fibers. Movement of sperm in the epididymis is a passive process as they are non-motile in the caput epididymis. Sperms are capable to be motile in cauda epididymis. Time taken by spermatozoa to cross epididymis in

Bull – 10 days

Stallion – 8 to 11 days

Ram – 13 to 15 days

Boar – 10-12 days

7. Vas Deferens

- It is lined with non-ciliated cells with little secretion. The lumen of vas deferens is lined with pseudostratified epithelium. It joins the epididymis to pelvic urethra. It acts as a temporary reservoir for sperm during their passage. The ductus deferens is held in place by the mesoductus deferens while running within the spermatic cord. The terminal part of the vas deferens is swollen (due to the abundance of ampullary glands) and called as ampulla which opens in the pelvic urethra via **colliculus seminalis** (a raised eminence).

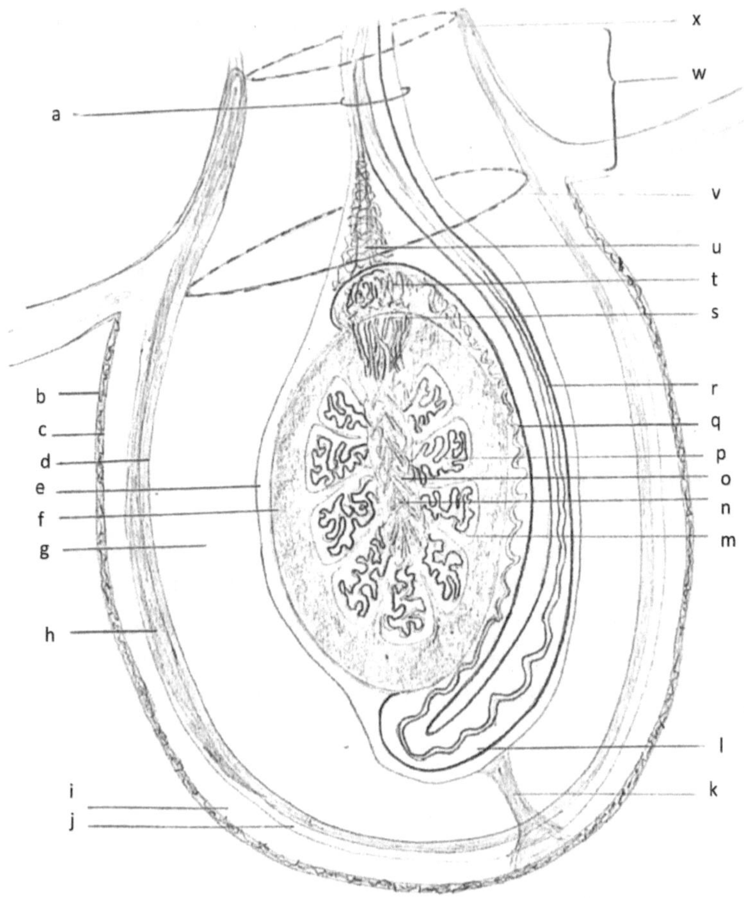

Schematic diagram of testis and surrounding structures

a. Spermatic cord; b. Scrotal skin; c. Tunica dartos; d. Tunica vaginalis communis; e. Tunica vaginalis propria; f. tunica albuginea; g. Vaginal cavity; h. Cremaster muscles; i. external spermatic fascia; j. internal spermatic fascia; k. scrotal ligament; l. Cauda epididymis; m. Lobules; n. Mediastinum; o. Rete testes; p. Seminiferous tubules; q. Corpus epididymis; r. Vas deferens; s. Vas efferens; t. Caput epididymis; u. Testicular blood vessels; v. Superficial inguinal ring; w. Inguinal canal; x. deep inguinal ring;

8. Ampulla

- Ampulla acts as temporary storage for spermatozoa. It is well developed in stallion but very small/absent in dog, tom, and boar. Ampulla acts as sperm reservoir in bull, ram, and dog.

- In the stallion ampulla and seminal vesicle secretes ergothioneine (an antioxidant).
- Uterus masculinus (utriculus prostaticus), a rudimentary male uterus, is the remnant of paramesonephric duct in male situated between the ampulla and vas deferens.

9. Seminal Vesicle

- It is the predominant vesicular accessory sex gland in bulls. It is situated near the neck of urinary bladder and lateral to the ampulla.
- It is lobulated and paired – bull and ram
  - Bag like – boar
  - Smooth and pear-shaped/pyriform – stallion
  - Absent – tom, dog, and camel
- Seminal vesicle has gland like nature so called as vesicular gland in most animals except in stallion where it is a more bladder-like structure and justified as a seminal vesicle.
- It contributes about 50% of total seminal plasma. Vesicular secretion in bull has **riboflavin** (yellow coloration of cattle semen). Seminal vesicle secretes citric acid, fructose, ascorbic acid, inositol, ergothioneine, riboflavin, KHCO3, Bovine seminal plasma protein (BSP), prostaglandin, etc.
- In boar; the seminal vesicle secretes **inositol** (osmoregulator) and **ergothioneine** (antioxidant). The gel portion of stallion semen is the secretion of the seminal vesicle.

10. Prostate Gland

- It is an unpaired gland. It has two parts pars propria and pars disseminate in the bull. These parts are joined with isthmus in stallion. In the ram, prostate has no body (pars propria). It is situated around the urethra at the neck of the urinary bladder.
- Secretion of the prostate gland has a lot of enzymes. The Prostate gland has acidic secretion (pH 6.5) with enzymes (proteinase, phosphatase, glycosidase, nucleases, nucleotidase, etc), anti-agglutinin, and Zn (in canines). In the dog prostatic secretion has citric acid, acid phosphatase, high zinc and chloride content, etc.

- The prostate is the sole (only) accessory sex gland found in the dog but less prominent in the tom. The size of this gland increases with age in the sexually intact male dog and the prostate pushes the freely movable urinary bladder from the pelvis into the abdomen. There is atrophy of the prostate when castration is performed as it is an androgen dependent organ.

11. Cowper's Gland (Bulbourethral gland)

- These are paired glands situated on both sides of the pelvic urethra at the level of the ischiatic arc embedded in the bulbospongiosus muscle.
- Its shape is –
  - Ovoid in bull, ram, and stallion
  - Cylindrical in boar (most prominent gland)
  - Absent in dogs
- The "**dribblings**" (urethra cleansing fluid) from prepuce before mounting in bull comes from cowper's and prostate gland. These secretions clear the urethra before semen ejaculation.
- In tom, the major accessory sex gland is the paired bulbourethral gland. The gel portion in boar semen which is meant for the vaginal plug/seal (due to the high content of sialomucin) after mating is the secretion of the cowper's gland.

- During ejaculation in the bull, ram, and doe there is a complete mixing of spermatozoa with fluids from the accessory glands. In stallion and boars, mixing is not as complete. Therefore, the ejaculates contain sperm-free and sperm rich fractions.
- Seminal plasma has potassium, sodium (max in boar) & calcium (max in bull), etc. It contains energy sources – fructose (in bull), sorbitol (in stallion), glycerylphosphocholine (GPC; in ram).

12. Urethra

- It has pelvic (on the pelvic floor), bulb or urethral isthmus (bending on ischiatic arch), and penile (in the penis) parts. It is a common passage for urine and semen.
- Urethral glands are present in boar but absent in bulls.
- Peristaltic action of the urethralis muscle assists in the transport of urine or semen through the pelvic urethra. Thus it aids in micturition and ejaculation.

13. Penis

- It is the muscular and cylindrical copulatory organ. It comprises two dorso lateral corpus cavernosum penis (CCP) and one ventrally situated corpus spongiosum (CSP) muscle. CCP is surrounded by tunica albuginea which is covered under the ischiocavernosus muscle which is called **erector penis muscles** (paired). Erector penis muscles exert pumping action on the bulbous portion responsible for penile erection. CSP surrounds the pelvic and penile urethra. CSP is covered by the bulbospongiosus muscle which helps in emptying the penile urethra at the time of ejaculation and urination. The root of the penis consists of two crura and the bulbus penis. The paired crura penis is the proximal end of the corpus cavernosum penis. This segment is attached to the ischiatic arch and is covered by the ischiocavernosus muscle. The bulbus penis is the caudal extent of the corpus spongiosum penis.
- Penis has an S-shaped curved muscular structure called as **sigmoid flexure** which maintains the desired length of the penis during intromission and ejaculation. It is
    - Post scrotal in bull and ram
    - Prescrotal in boar and camel
    - Absent in stallion
- **Retractor penis muscles** (paired) are smooth muscles, attached from the dorsal apical ligament of the penis to sacrum or $1^{st}/2^{nd}$ coccygeal vertebra. This muscle is meant to draw/retract back the penis in the prepucial sheath after ejaculation.
- Penis may be :
    1. **fibroelastic** type; characterized by lesser erectile tissue and more fibroelastic tissue with less blood accumulation due to small cavernous space e.g. bull, ram, camel, and boar. Here penis achieves its length due to the straightening of sigmoid flexure.
    2. **Musculo vascular** type; characterized by more erectile tissue with more blood accumulation due to bigger cavernous space which is mainly responsible for lengthening of the penis e.g. stallion, dog, tom. Here penis achieves its length due to the hydrostatic pressure of blood in the cavernous space (Sigmoid flexure is absent).
- Os penis, a bone in penis, is found in the dog. The spiral penis is the character of boar.

- Length of the penis –
    - Bull – 36 inches
    - Ram – 30 cm
    - Stallion – 50 cm
    - Boar – 50-55 cm
- Erectile tissues are more in stallion than bull penis but retractor penis muscles are stronger in bull than stallion. The urethral process, corona glandis, fossa glandis, prepucial ring, and urethral sinus/diverticulum are characters in the stallion penis. In stallion, the projected end of the urethra outside the glans is called the urethral process which is situated at the fossa glandis. Smegma is infectious material filled in the urethral sinus in stallion and may cause infertility in females due to natural covering. The urethral process is also a characteristic of the ram penis.
- The terminal or apical part of penis is called as **glans penis**. Glans penis is absent in boar and tom. The Glans penis in dog has two parts: Bulbus glandis and pars longa glandis. Both have erectile tissues. The cranial portion of the dog penis has a bone called as ospenis/baculum. Its ventral groove contains the urethra. Os penis allows the dog to achieve intromission before complete penile erection.
- In the dog, during penile erection, the pressure increases in the corpus spongiosum first. Engorgement of the corpus cavernosum is delayed, but there is three times more pressure than the corpus spongiosum. Intracavernosal resistance decreases due to the relaxation of smooth muscles in the corpus cavernosum with an increase in arterial blood flow and there is occlusion of venous outflow by obstruction at the tunica albuginea and compression by the ischio-urethralis muscle. Penis erects due to this increased hydrostatic pressure because of blood pooling. The penis of male dogs twists 180 degrees during the copulatory lock and this twist may assist with venous occlusion and maintenance of erection during the copulatory lock, which is unique to dogs. Detumescence or normalization of the penis occurs due to reduced hydrostatic blood pressure in the penis. This occurs by (1) Sympathetic innervation via the hypogastric nerve increases arterial resistance, causing a decrease in corpus cavernosal pressure and (2) smooth muscle contracts in cavernosal tissue, permitting the widening of venous lumen and restoring venous outflow.

- Tom's penis points posteriorly in the relaxed stage. In tom, there is no bulbus glandis, no glans penis, no/short os penis. At the terminal part of the penis, there are several keratinized papillae/spines (100-120) which are responsible for the "**cat cry**" during withdrawal of the penis after mating. These spines are developed fully at puberty and stimulate luteinizing hormone (LH) release in female at the time of coitus.
- In bull, there are prepucial muscles called as protractor and retractor prepucial muscles which are meant for drawing prepucial opening forward or backward. The stallion has a double folded prepuce. Sometimes in the outer fold, there is some wax-like secretion called as "**beans**". The boar has a preputial diverticulum (pouch) dorsal to the prepucial orifice, which contains decomposed urine and dead epithelial cells/debris. The fluid in the diverticulum also contains a pheromone that is responsible for restricting the sow movement during coitus.

14. Prepuce

- It is invaginated fold of skin over the penis and penile integument (modified skin covering the apex of penis) which is called as prepuce. It is homologous to labia majora in females. The prepucial orifice or **prepucial ostium** is the opening of prepuce which is 6-7 cm behind the umbilicus. The orifice of the prepuce is surrounded by long and tough preputial hairs in the bull.
- Swine and ruminants urinate inside the prepuce, whereas horses, dogs, and tom extend the penis beyond the prepucial sheath while micturition.

Blood supply – Testes – Internal spermatic artery
Scrotum – external pudendal artery
Penis – External and internal pudendal artery with obturator artery
Accessory sex glands – Internal pudendal artery

Nerve supply – Testes – Renal and mesenteric plexus
Scrotum – Genital nerve (from the genito-femoral nerve)
Penis – Pelvic plexus, pudendal & hemorrhoidal nerve
Accessory sex gland – Hypogastric nerve (sympathetic innervation) and pelvic plexus

a. Penile structures of different species:

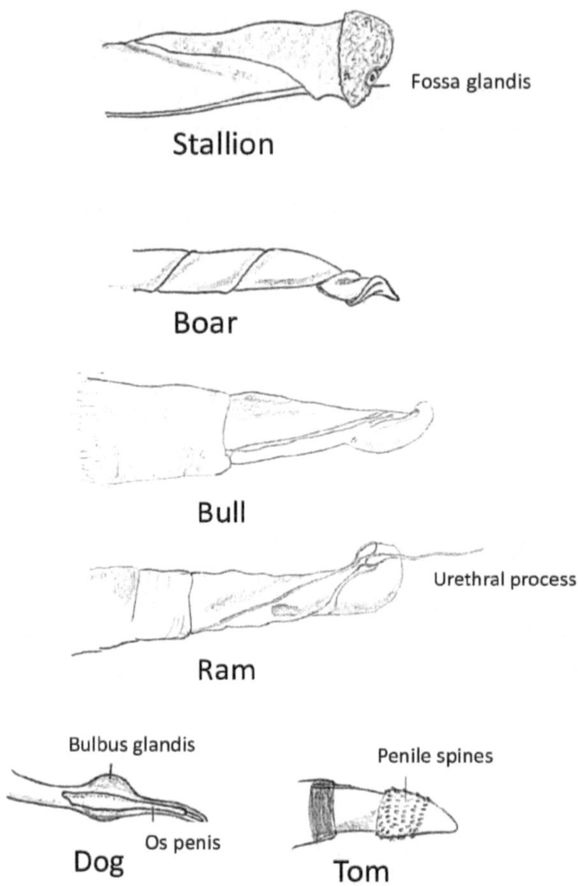

Penile structures of different species

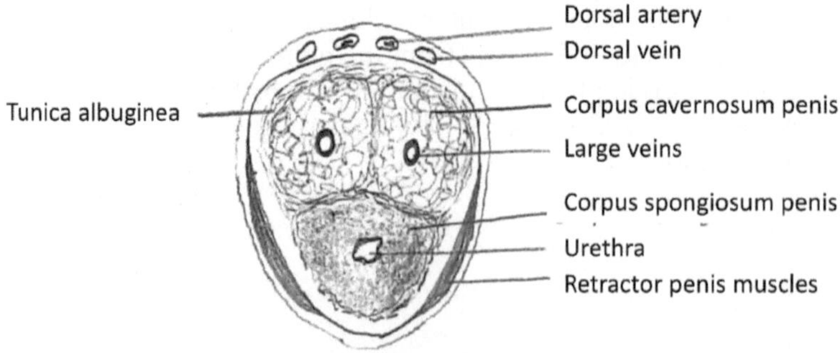

Cross section of bull penis

Thermoregulation of the Testes

- Optimum sperm production or spermatogenesis requires 2^0-5^0 C less temperature than body core temperature.
- Various structures play a role in thermoregulation process of testes as:
  1. Thermoreceptors are present on the scrotal skin which regulate scrotal temperature through sweating and panting like reactions according to environmental temperature.
  2. Adrenergic sweat glands are present on scrotal skin.
  3. There is absence of subcutaneous fat in scrotal skin for better heat dissipation.
  4. Contraction and relaxation of tunica dartos muscles alter the thickness and surface area of the scrotal wall. In hot environment tunica dartos becomes thin and relaxed while in cold environment it becomes thick and wrinkled.
  5. In hot environment, the cremaster muscle relaxes due to which testicles move away from the body and in cold environment, it contracts due to which testicles come closer to the body.
  6. The close apposition of the artery and veins allows heat exchange to occur between spermatic artery and vein, so that the temperature in the testis remains several degrees lower than the core body temperature i.e. Testicular arterial blood becomes cooler due to the coiling of spermatic vein (Pampiniform plexus) over spermatic artery.
- The length of the spermatic artery (approx 5 m) results in the arterial pulse being almost eliminated. It is considered that pulsatile blood flow is not supportive for spermatogenesis.
- In boar, there is less pendulous scrotum and less efficient sweating. In boar, there is a difference of about 3.2^0 C between rectum and scrotum temperature.
- In ram, the temperature difference between spermatic artery and vein is about 4^0 C.

2

Endocrinology of the Male Reproduction

Chetna Gangwar and Dileep Kumar Yadav

- In male animals, endocrine regulation of reproduction is mainly controlled by pituitary gonadotrophins and gonadal androgens. It is not the same as in females as there is ***only negative feedback controls in case of male*** while it is positively as well as negatively feedback control in females. In other words, the synchrony of pituitary gonadotrophins with gonads in male is non cyclic while in female it is cyclic i.e. even after castration (absence of androgen) the pulse frequency of LH and FSH is retained.
- Sex Hormones are sex specific e.g. estrogen and progesterone work primarily in females and testosterone in males but sex specificity is limited as far as sexual behaviour is concerned. As estrogen injection in castrated ram recovers the maleness while testosterone injection in ovariectomized female maintain the receptivity for male up to a certain extent.
- Hypothalamus acts as the interface between nervous system and endocrine system. GnRH is synthesized in neurons in hypothalamus, released at nerve endings in an episodic episodic manner and transferred via hypophyseal portal system to anterior pituitary to stimulate the secretion of gonadotrophins.
- Male reproduction is mainly controlled by androgens (Testosterone) secreted by gonads or teticles under the influence of pituitary gonadotrophins (LH / ICSH and FSH). Pituitary gonadotrophins are synthesized and secreted in response to the hypothalamic gonadotrophin releasing hormones (GnRH).
- LH receptors are found on the leydig cells and germ cells. Testosterone receptors are found on the leydig cells, sertoli cells and myoid cells but not on the germ cells.

- LH receptors are less in numbers and less sensitive on leydig cells up to the age of 6-7 months of age i.e. up to this age LH (ICSH) secretion does not stimulate the secretion or rise in the level of testosterone. Growth hormone, insulin and insulin-like growth factor (IGF)-1 have stimulatory effect on the action of LH/ICSH over LH receptors.
- Acting on the leydig cells, LH (ICSH; Interstitial cell stimulating hormone) regulates steroidogenesis. LH controls the transformation of cholesterol into pregnenolone (rate-limiting step of testosterone steroidogenesis). Pregnenolone converts into progesterone and further into testosterone (18 C) in leydig cells.
- LH (ICSH) is secreted in a pulsatile fashion (Basal value and peak value are 5ng/ml and 15-30 ng/ml in bulls). Episodes of LH secretion from the anterior pituitary occur after each 2-4 hr under the influence of hypothalamic GnRH (4-10 peaks of LH per day). LH acts through LH receptors found on the leydig cells and leads to testosterone synthesis (Basal value 2-8 ng/ml). After 30-60 minutes of LH pulse, the level of testosterone is at peak (10-20 ng/ml) which becomes basal again after 40-80 minutes. The increased level of testosterone imposes negative feedback on hypothalamus as well as pituitary.
- In dogs, LH basal level is 1.0-1.2 ng/ml and peak is 3.8-10.0 ng/ml and in response the level of testosterone increases from 0.5-1.5 ng/ml (basal level) to 3.5 -6.0 ng/ml. In tom, the concentration of LH ranges from 3 to 29 ng/ml and in response the concentration of testosterone ranges from 0.1 to 3.3 ng/ml.
- Synthesized androgen (testosterone) converts into DHT (5α - dihydrotestosterone) in sertoli cells and accessory sex glands by 5α-reductase enzyme. Testosterone converts into estradiol in sertoli cells by aromatase enzyme. This aromatization also occurs in adipose tissues and hypothalamic pre-optic area. A high level of blood estrogen is an indicative of sertoli cell tumour and occurs mainly in canines. In stallion, the maximum level of estrogen (produced from testosterone) is found in epididymis of stallion as seminiferous tubule and epididymis produce a high level of testosterone.
- There is negative feedback response of both of these products (DHT and estradiol) at the level of hypothalamus and pituitary for LH (ICSH) secretion.
- DHT is more potent androgen than testosterone as DHT does not undergo aromtization. DHT mainly controls accessory sex gland activity whereas testosterone is the primary androgen involved in spermatogenesis.

- Both testosterone and DHT are bound by androgen-binding protein (ABP) within the reproductive tract lumen and maintain high androgen concentrations in the seminiferous tubules and epididymis. Maximum testosterone is transported in the blood in bound form with α-globulin and some form is free which is converted into DHT.
- FSH (follicle stimulating hormone) is secreted in non episodic pattern. FSH regulates the proliferation of sertoli cells in prepubertal life i.e. determine the sertoli cell numbers at puberty. It determines the postpubertal spermatogenic yield as sertoli cells regulate the process of spermatogenesis. FSH is responsible for the initiation of spermatogenesis but FSH and testosterone both are required for the maintenance of spermatogenesis process and spermiation.
- **Sertoli cells** are also known as nursing mother cells or sustentacular cells. They regulate spermatogenesis and nourish the spermatozoa. The functions of the sertoli cell are under control of FSH. Under control of FSH, sertoli cells secrete :
    1. **Androgen binding protein** (ABP) – It binds the androgen (testosterone and DHT) and maintains their concentration in seminiferous tubules and epididymis. Steroid binding protein in males is α-globulin.
    2. **Activin** – It regulates the development of leydig cells in fetal testes and at puberty. It is also responsible for the delay in growth of leydig cells till puberty. It promotes the secretion of FSH at the pituitary level.
    3. **Inhibin** – It regulates the various differentiating and growth factor functions. It regulates leydig cell function also. It also exerts negative feedback effect on the secretion of FSH at pituitary level. Inhibin B is more potent and predominant in most of the species whereas in Ram, inhibin A is found to be more potent.
    4. **Anti-mullerian hormone** (AMH) or Mullerian inhibiting hormone (MIH) – It inhibits the development and differentiation of the mullerian duct system or paramesonephric duct system in embryonic life i.e. responsible for the development of male fetus.
    5. **Transferrin** – Source of iron for sperm.
    6. **Glutathion** – Acts as anti-oxidant
    7. Insulin like growth factor (IGF-I)

- Prolactin receptors are present on leydig cells. This hormone promotes the functions of LH receptor on leydig cells. But a high level of prolactin, growth hormone (GH/STH) and melatonin exerts negative feedback action for production and release of gonadotrophins i.e. responsible for delayed development of sexual characters and sexual behaviour (delayed puberty and libido).

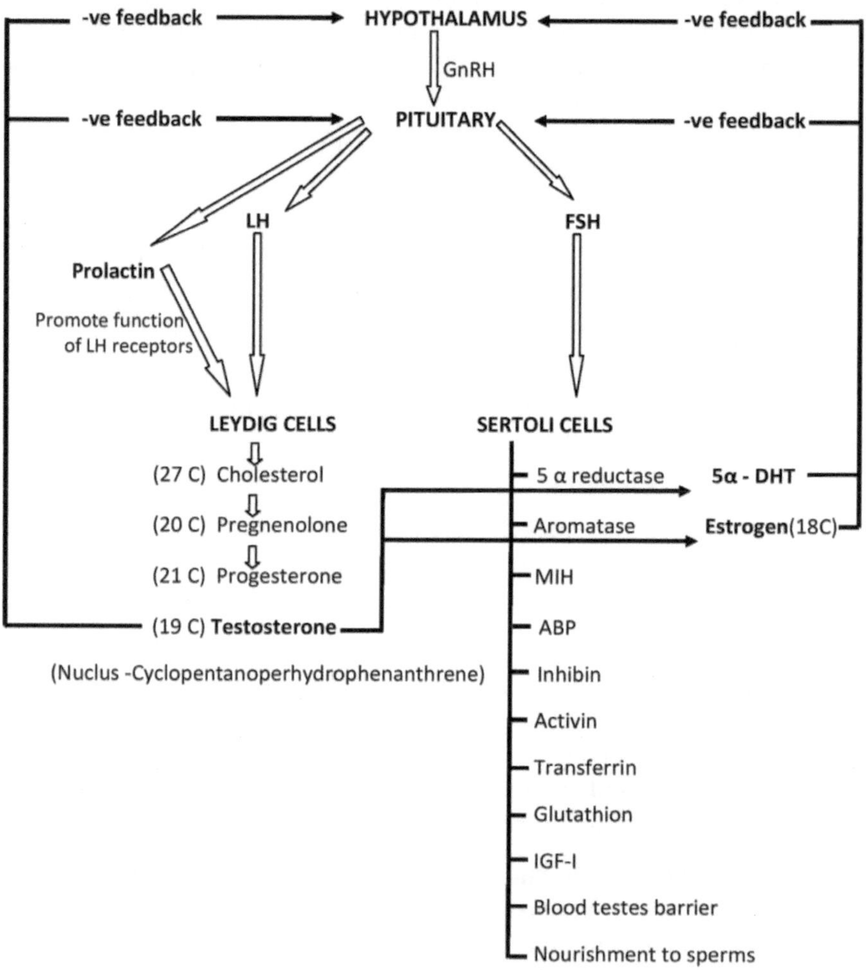

Endocrine regulation of male reproductive system

Functions of the Androgen

1. Testicular descend
2. Sexual differentiation of male reproductive system

3. Regulation of structure and function of accessory sex glands and ductus system
4. Spermatogenesis
5. Development of secondary sexual characters and maleness (Body confirmation, hairs, hump, broad shoulder, aggressiveness, muscle growth etc)
6. Prolongation of epididymal sperm life
7. Different sexual behaviour patterns in different species
8. Penile and prepucial growth. Degeneration of penile frenulum
9. Urinary pattern in dogs (urination with one lifted leg)

- Prepubertal castration leads to the absence of libido, underdeveloped genitalia and no mating behaviour. Post pubertal castration leads to stoppage of ejaculation first then mounting behaviour and then penile erection stop.
- In Dogs, there is hypertrophy and hyperplasia of prostate gland due to over-production of testosterone (Because of overstimulation of leydig cells by gonadotrophins).
- Thyroid stimulating hormone (TSH), Growth hormone (STH), Adrenocorticotrophic hormone and Melatonin are necessarily required for maintenance of male reproduction but their over secretion may lead to reverse or negative feedback effect.
- Oxytocin also plays a role in increasing tubal activity at the time of ejaculation leads to increased seminal volume with more sperm concentration.
- Some endogenous opioid peptides like β-endorphin are present in testicular interstitial fluid in high volume and participate in leydig cell steroidogenesis.
- stallions and rams are seasonal breeders but they are not azoospermic during the non breeding season but there is significant reduction in sperm production.

Endocrine control in seasonal breeder

- Stallion is reproductively more active during long days. During short days, there is less light duration stimulates the pineal gland via retina and optic nerve to secrete more melatonin which further inhibits the production and secretion of GnRH and so the gonadotrophins and less steroid (like androgen) production. But in long days, there is more gonadotrophin and steroid formation occurs due to less melatonin production. Camel is also a short day breeder. Occurrence of long days after short set a rhythm which is found to be necessary for stallion reproduction. Without a period of 2 to 3 months of short day followed by long day will decline the libido as well as sperm production in stallion.

- Ram is more reproductively active during short days. Unlike stallion, due to the short day length the increased melatonin secretion reaches to hypothalamus and stimulates the pulsatile secretion of GnRH which in turn is responsible for more steroid production. Reducing the daylight hours in ram reduces the prolactin production. It leads to increased gonadotrophins secretion and hence testosterone.

3

Evaluation of Breeding Soundness in Male Animal

Vikas Sachan and Avneesh Singh

- To identity the normal fertile, infertile (Temporary loss of fertility), subfertile (reduced fertility), or sterile (no fertility) male animals makes the requirement of breeding soundness evaluation necessary. Infertile and sterile animals prove to be harmful for the economic survival of the dairy industry and should be eliminated from the breeding management system.
- The breeding soundness examination is very necessary for the breeding management system as the ultimate goal is the achievement of healthy pregnancies. It is also useful for successful artificial insemination (AI) programs that involve the collection and cryo-processing of quality semen.
- It involves the examination of management, reproductive anatomy, physiology, or any disease condition which affects reproductive performance.
- Males, detected as infertile or subfertile, are needed to be re-examined after some time because they may recover or gain maturity toward fertility in due course of time.

1. History and General Physical Examination

- The breeding history of previous breeding performance in terms of libido, reaction time, and pregnancies along with breeding records related to no. of females bred, over/underuse of sire, male to female ratio, mating frequency, total females bred, etc. should be thoroughly considered.
- Management (housing, feeding), and vaccination programme, etc. also reflect the breeding soundness status.

- History related to matings/inseminations and temporary loss of fertility in past helps to determine infertility causing (congenital or acquired) conditions.
- History of abortion or post-insemination uterine infection in females bred by the sire should also be considered for the possibilities of genital infections & venereal diseases.
- History related to the occurrence, duration, and severity of any systemic or local disease (especially involving testicles) condition or inflammation within a period of the last 3-4 months (1-2 spermatogenesis waves) should be considered during the evaluation of breeding soundness of any male animal.
- Age is an important factor to be considered. Older age animals have reduced libido and ability to copulate because of impaired hormonal balance in the body and also due to increasing bone and joint problems with age. Seminal quality is also decreasing with increasing age. Very young bulls are also questionable in terms of sexual maturity achievement and sexual behaviour. Very young bulls have smaller testes with short epididymis i.e. sperm output is less with more immature sperms. In younger bulls, there are more numbers of morphologically abnormal spermatozoa, so re-examination is needed after some time with more ejaculates.
- The puberty directly relates to body growth and confirmation (thin & weak body with poor nutrition and lethargic, fatty body with heavy scrotal fat in case of very high plane of nutrition etc). which directly affects the breeding ability. Bulls having body growth rate $\geq$ 1.5 kg/day are more prone to joint diseases, obesity, scrotal fat deposition, impaired spermatogenesis, and poor libido.
- The body weight and size of the sire should be appropriate according to the size of females to be copulated. Generally, females get injured due to overweight and heavy-sized bulls and boars.
- From a distance when a bull is in a small free area, observation related to gait, posture, and any gross abnormality or abnormal body confirmation should be taken.
- Close observation (in a restrained bull) related to body organ anatomy should be carefully done from the front & rear and both sides (left & right) of the body. Bulls with cow hock, sickle hock, bow-legged, spastic paresis, crampiness, honeymoon back, etc. are not suitable for breeding purposes.

- Neuro-muscular and joint confirmation are of prime importance and directly related to breeding soundness examination. Organs related to hearing, sight, and olfaction should be examined carefully.
- Physiological parameters such as rectal temperature, respiration rate, and heart rate should be normal. Palpation and auscultation of vital organs should be performed carefully to rule out any abnormality. Ultrasonography (USG), radiography, or cytology may be performed, if needed.
- Pain symptoms especially related to the hind limb and vertebral column are directly related to the inability to copulate i.e. indicative of poor breeding ability. Corticosteroids produced due to pain stress may impair normal spermatogenesis.
- Very aggressive temperament or very lethargic males are not considered to be fit for breeding purposes.

2. Genitalia Examination

- According to the temperament of the male, it should be restrained or sedated properly. Palpation and ultrasonography examination are useful tools to diagnose any abnormal swelling of genitalia (inflammation, neoplasia, adhesion etc.), abscess, calcification, or fibrosis etc.
- Testes should be firm or slightly soft, resilient, and freely movable in the scrotum without any signs of adhesion. Atrophic, spongy, flaccid, or flabbiness in testes may be an indicator of testicular degeneration and such animals are not used for breeding. Extremely hard testes or an area of the testicle with irregular contour is indicative of chronic testicular degeneration with fibrosis or calcification. An extremely soft/spongy testicle or part of the testicle indicates an abscess or degeneration. Testes should be examined carefully for cryptorchidism, Orchitis, tumour, hypoplasia, or signs of degeneration.
- Scrotal circumference and hence semen volume are directly related to daily sperm output. The scrotal circumference can be measured with a tension tape. Trans-scrotal USG or callipers can be used for the same in a stallion and boar. It should be 26-28 cm at puberty and 32-36 cm in adult cattle bull. The scrotal circumference should be 24 to 28 cm in buffalo bull and 28-34 cm in ram. Scrotal circumference may be lesser in the non-breeding season (25-35 % smaller) in ram.
- Bulls with short scrotum (poor thermoregulation) or very large scrotum (chances of traumatization) are not recommended for breeding.

The scrotum should be thin and cooler without any signs of trauma, laceration, inflammation, or parasites (e.g., mange) normally. Soft/spongy or flaccid head or tail of epididymis (lost turgidity) indicates the depleted sperm reservoir or disruption of the part. Spermatic cord (at the level of the inguinal ring) palpation is recommended to rule out any adhesion, fibrosis, scrotal hernia, varicocele, or any other abnormal structure.

- The vesicular gland and ampullae are palpated (rectal palpation) in bull and stallion for any pathological finding (pain, swelling, adhesion, fibrosis). Vesicular gland pathologies are more common in bull while abnormality of ampullae is more common in stallion. Prostatic pathologies can be better examined by radiograph or ultrasonography than digital palpation in smaller boar and dogs.
- Examination/palpation of the penis and prepuce for any swelling, adhesion inflammation, trauma, scalds, scabies, neoplasia, abnormal discharge, or stenosis, etc. should be done. Excess pendulous sheath makes prepuce prone to prolapse. Penile abnormalities (persistent frenulum, deviation, etc.) can be diagnosed by observing the normal erection, mounting, intromission, and ejaculation during service to female or semen collection.
- Alkaline phosphatase (ALP) is produced in the epididymis. In an azoospermic semen sample, measurement of ALP is essential to determining weather the cause of azoospermia is poor libido/testicular failure, or ductal blockage. In an azoospermic sample, if the ALP is low, it indicates the ductal blockage, whereas a normal ALP indicates testicular failure.

3. Mating Behaviour

- Male animal with good libido may mount on female (in estrous or not in estrous), on male or dummy as well according to the training given. If the animal is unwilling to mount then past bad experiences, stressful conditions (e.g. recent transportation), the presence of a senior male, and the presence of a strange person nearby, etc. should be taken into consideration before judging him for breeding soundness.
- The **libido test** predicts a bull's performance based on his urge and no. of mating attempts on a restrained estrous female.
- The **serving capacity** test predicts the mating ability based on a number of effective services when 4-6 bulls are admitted to restrained estrous females.

- **Service test** is a combination of libido test and serving capacity test. It predicts the quantitative and qualitative aspects of the bull's performance. Bull with moderate no. of mounts and higher pregnancy rate is considered to have good breeding soundness. It is not a good predictor for bulls who have never mated or at their first breeding season.
- Determining the pregnancy along with the ratio of mating to intromission for inexperienced & young bulls and number of mounts for older bulls is a good criterion to predict the breeding soundness.

4. Seminal Examination

- Semen quality examinations provide essential information about the fertility status of the animal, but they do not wholly define the fertilizing capacity of the animal due to the involvement of several other factors along with female factors.
- *Gross examination* like seminal volume, colour, consistency, density, odour, pH, or any abnormality indicates the seminal quality at the initial stage to help in determining the breeding soundness of the bull. These examinations should be completed within 8-10 minutes.
- *Microscopic examinations* like mass motility, progressive motility, sperm concentration, % live spermatozoa, sperm morphological abnormality, % intact acrosome sperms, presence of other cells, and sperm membrane integrity, etc are good predictors for seminal quality.
- *Biochemical examination* related to the metabolic activity of spermatozoa e.g. fructolysis index, MBRT, RRT, Buffering capacity etc and enzymatic activity are also predict the spermatozoa quality in a sample.
- *Fertility tests* like Zona binding assay, Hemizona assay, Zona free hamster ova penetration test and Bovine cervical mucus penetration test determine the ability of sperm cells to attach to zona pellucid as well as their penetration.
- *Several other tests* like Acrosomal integrity determination, DNA fragmentation index (Test for chromatin integrity), Antisperm antibody test etc are also performed to judge the sperm quality more accurately.

4

Puberty and Sexual Maturity

Atul Saxena and Vikas Sachan

Puberty

- It is the age at which sexual organs are functionally developed and gonads start responding to GnRH to start steroidogenesis and spermatogenesis. Animal at puberty develops secondary sexual characters with proper libido.
- At puberty, the bull is able to produce 50 million spermatozoa with 10% progressive motility in 1st seminal ejaculates. It should be 100 million spermatozoa with 10% progressive motility in 1st seminal ejaculate in case of stallion.
- At puberty, gonadotrophic cells are functionally developed in the mature pituitary tends to gonadal maturation. The steroidogenesis starts properly which leads to the development of genital organs and secondary sexual characters.
- Due to reduced brain steroid receptor sensitivity and increased gonadotrophins secretion, there is activation of spermatogenesis at puberty. In the prepubertal period, testosterone crosses the blood brain barrier and is converted into oestradiol by aromatase enzyme. Oestradiol and testosterone exerts inhibitory effect on hypothalamic GnRH neurons which leads to reduced secretion of gonadotrophins. Also, during the growth phase, STH (somatotropic or growth hormone) increases which results in reduced production of gonadotrophins and thus no spermatogenesis occurs. But at the onset of puberty, the sensitivity of hypothalamus for the inhibitory effect (negative feedback effect) of oestradiol and testosterone is reduced and thus increases pulsatile secretion of GnRH. It results in increases secretion of FSH & LH which stimulate spermatogenesis in testes. Also, when body growth becomes steady (around puberty), there is a reduction in STH secretion. It also results in increased secretion of interstitial cell stimulating hormone

(ICSH), which in turn increases the testosterone secretion from gonads. Gonadotrophins and testosterone are responsible for the development of libido and attaining puberty in male animals.

- Testicular circumference is more accurate predictor of puberty than body weight and age. It should be 25 to 27 cm at the age of puberty in bulls.
- Age of puberty : Cattle bull (exotic) - 1.0 to 1.5 yrs

Cattle bull (indigenous)	- 1.5 to 2.0 yrs
Buffalo bulls	- 14 to 16 months
Stallion	- 1.5 to 2.0 yrs
Camel	- 3 yrs.
Ram	- 7.0 to 8.0 months
Boar	- 6.0 to 7.0 months
Dog and tom	- 8.0 to 10 months

Sexual Maturity

- It is the age at which there is complete physiological (behavioural & spermatogenic) reproductive development of male animal which results in fertile services with normal reproductive capacity.
- In the case of stallion, it should be considered as potential breeder if he is able to produce 400 million sperm cells in his first ejaculate in breeding season with 50% motility (provided 60% normal sperm morphology). In the case of boar, it should be able to produce 1000 million sperm cells with 60% motility and in the case of ram, it should be able to produce 1000 to 1800 million sperm cells per ml semen with 60% motility.
- The period between puberty and sexual maturity is called "**adolescence**".
- After the first mounting of a male animal, there is faster development of penile structure. Its shape and size and detachment of penile frenulum occurs which makes animal ready for its first ejaculate after some days.
- From the age of 2 months in bull to 6 months, there is poor sensitivity of leydig cells to ICSH / LH. But from 6 months the sensitivity of leydig cells to LH increases gradually due to which androgen (1st androstenedione and later on testosterone) concentration in blood increases and animal attains puberty at a particular age.

- In buffalo bulls active spermatogenesis can be observed at 14 to 16 months of age and they become sexually mature at the age of 2 to 2.5 years.

Factors Affecting Puberty and Sexual Maturity

1. The age of puberty depends not only on the genetic makeup of an individual as well as on the environment in which they live.
2. Age of puberty varies according to species, breed, strain and individual. Puberty attainment is earlier in exotic bulls than in zebu cattle as well as earlier in cross bred than inbred animals.
3. Body weight or physiological growth is more closely related to puberty than age. The body weight of cattle and buffalo bull should be 240 and 280 kg respectively at puberty. At puberty, body weight of ram should be 60% of adult weight while it should be 45% in the case of cattle bull.
4. TDN, protein, minerals and vitamins affect the age of puberty. Nutritional status regulates not only growth rate and body weight but also the endocrine balance e.g. gonadotrophins from the pituitary which directly relates to puberty and sexual maturity.
5. Obesity, letharginess and laziness due to high plane of nutrition delay puberty. High plane of nutrition causes the deposition of fat over gonads and genital organs which hampers the thermoregulation of testes. Due to fat around the genitalia, there is poor secretion of reproductive hormonal and absorption of some hormone by the fat lead to a reduced concentration of reproductive hormones in blood which affect puberty and sexual maturity.
6. Poor plane of nutrition leads to deficiency of various vitamins and minerals which delays puberty due to poor body growth and abnormal endocrine balance. Deficiency of Vit A and Vit E causes pituitary cyst and may cause irreversible seminiferous tubule degeneration and testicular degeneration which badly affect puberty and libido. High-molybdenum and low-copper diets have resulted in low libido and degeneration of the seminiferous tubules and testicular interstitial tissue thus affect puberty.
7. Improper balance between gonadotrophins and androgen delays the age of puberty. Melatonin from the pineal gland increases brain steroid receptors sensitivity which delays puberty.
8. Some social factors like male reared in isolation have delayed puberty with poor libido. Presence of the female in herd hasten the puberty of male animals. The onset of boar contact at a younger age decreases age at puberty but requires a longer period of stimulation.

9. Male animals reared in stressful environment such as extreme cold, hot and humid have poor sexual activity. Reducing the daylight hours in ram reduce the prolactin production. It leads to increased gonadotrophins secretion and hence testosterone.

5

Libido

Jitendra Kumar Agrawal and Vikas Sachan

Libido

- It is the willingness & eagerness of male animal to mount and perform complete service. It is sexual desire or drive which is an innate character of male animals.
- Libido and mating ability have no direct relation with fertility. Libido is not related directly to sperm production and semen quality.
- The libido of an animal depends on androgenic steroid hormone. Vasectomy or cryptorchidism does not affect libido in male animals.
- Animals have no or reduced libido that are castrated before puberty. If the animals, that know the copulation, are castrated after puberty, they display the copulation behaviour for longer period.
- **Reaction time** – It is the time between the contact of a male animal with female animal to 1st false mount with erected penis. Reaction time is the good predictor of libido.
- For bulls, Reaction time should be less than 3 minutes.

≤ 1/2 min	Excellent libido
1/2 to 1 min	Good libido
1 to 3 min	Fair libido
≥ 3 min	Poor libido
> 15 min	Problematic

- Assessment of libido and serving ability is predicted by the "**serving capacity test**" (comparative mating test according to number of mounts and effective services) and "**libido test**" (individual bull performance according to numbers and eagerness of mating attempts). Combining these two tests, there is one single test called "**service test**" (Qualitative and quantitative estimation of bulls mating performance). It is based on effective mounts (intromission and ejaculation), total mounts and also their ratio. The score of service test depends on and changes according to rearing condition, sexual experience, exposure to estrous female and breeding season etc.

- Masturbation is considered normal for almost all stallions and it does not affect the libido. There is protrusion, erection of the penis and rubbing against the abdomen but ejaculation rarely occurs. It may occur after every 1.5 to 2 hours interval.

Factor Affecting Libido

1. **Heredity and genetics**– Offspring of good libido animals have good libido. Twins have same libido. Generally drought breeds have better libido than milch breeds. Exotic bulls have better libido than Indigenous.
2. **Nutrition** – TDN, protein, macro and micro minerals (Ca, P, Cu, Fe, Co. Mn etc) and vitamins are required to maintain optimum libido. P, Co and Vit A are very important for maintaining libido. High-energy diets may reduce bull libido. Over feeding of protein-rich diet may not increase the reproductive performance of the male besides it may deteriorate the seminal quality. Deficiency of minerals and vitamins leads to poor endocrine balance in the body and hence affects libido. Overfeeding makes animal obese and lazy which may lead to joint problems (legs and vertebral column) and reduce the libido. Enlarged rumen due to excess roughage feeding makes copulation difficult and reduces libido. High-molybdenum, low-copper diets result into low libido. Zinc appears to be of primary importance to adequate reproductive function in the male. Deficiency of Zn may result in retarded testicular growth, atrophy of tubular epithelium, reduced pituitary gonadotropin output, and reduced androgen production. Iodine deficiency is associated with depressed libido and deterioration of semen quality. Cobalt deficiency, hence cobalamin deficiency, is associated with anaemia and reduced libido.
3. **Senility** – In younger bulls, libido may be poor with delayed puberty because of poor feeding, low nutritional status and inexperience. In over aged bulls, libido may be poor due to low testosterone, excessive use and joint problems like arthritis.
4. **Management** – Training and handling of male animals directly affect the libido. Changing the site of semen collection, time of semen

collection, unsuccessful attempts of copulation by immature bull to large sized female, changing the bull attendant, harsh and abusive handling and frequent punishments affect the libido of animal psychologically. Improper footing, overgrown hooves, discomfort and a stressful environment reduce the libido. Poor exercise, isolation, frequent changes in the environment and the presence of other older and experienced bulls frighten the young bulls. Fighting, kicking and application of too hot or cold artificial vagina is responsible for a bad experience for the bull which leads to poor libido.

5. Very **frequent ejaculation** leads to reduced libido. **Changing female** may restores the libido in males.
6. **Climate** – Libido in hot and humid climate and summer season (> 40^0C) is poor. Generally climatic factor affects libido in buffalo bulls, exotic bulls and rams while it affects less in native bulls and rams.
7. **Endocrine factor** – Poor endocrine metabolism leads to poor androgen level and its sensitivity. Testosterone to estrogen ratio is more important in regulating libido than their individual level. Poor testosterone to estrogen level represents poor libido. Hypothyroidism, hypopituitarism and hypogonadism reduce the libido.
8. **Injuries and diseases** – Infection and inflammation of male genitalia directly affect the libido especially when there is a painful condition of the penis and prepuce as balanitis and posthitis respectively. Injuries to joints, muscles, tendons and nerves especially of limbs and vertebral column reduce the libido. Any systemic illness such as pneumonia, enteriritis, tuberculosis, gastritis, peritonitis etc causes pain, fever, anorexia and weight loss which is directly or indirectly responsible for reduced libido. In stallion mounting or thrusting may be affected by painful conditions like cauda equina neuritis, ilio-femoral thrombosis, laminitis, arthritis, and coital exanthema. Poor libido in tom may be due to congenital and acquired (penile hair ring) phimosis condition. Penile hair rings may lead to unsuccessful ejaculation even after long bouts of unsuccessful pelvic thrusting.

6

Sexual Behaviour in Male Animal

Anuj Kumar and Vikas Sachan

- Sexual activity or mating behaviour is innate character of the males. Unlike the female animal, male animal has a nature of sexual receptivity permanently.
- Male gonadal steroid androgen controls sexual behaviour. At regular interval of 2-4 hrs, there is secretion of testosterone which converts into estradiol due to aromatization. Due to this very frequent rise in estradiol level, male shows sexual behaviour permanently. In female animals, a rise in estradiol occurs only during estrous.
- When in a group of animals, any female can be mated with any male, it is called **promiscuous sexual behaviour**. Sexual behaviour in some animals is according to their social structure and territorial/home range i.e. male and female share some of their home range for the purpose of mating e.g. Roe deer, Muntjak antelope etc. In some animals, both males and females share whole of their territory for the mating purpose called a **harem** e.g. wild rabbit. Some animals come into a temporary territory meant only for mating in breeding season e.g. seal.
- Rut is a term used to describe breeding season in animals like deer, sheep, camel, goat, bison, giraffe, antelopes, elephant etc i.e. period of increased testosterone level in the blood with increased interest to females. The rutting period or mating season in sheep is called "**tupping**". Male camel has poll glands at the back of the neck near the poll. These are subcutaneous exocrine glands. They are activated during the rutting season after sexual maturity. Their dark brown secretion has pheromones that are attractive to females in the breeding season.
- The Sexual pattern/behaviour of a species is determined and influenced according to not only their genetic makeup but also as per the physical and social environment they are exposed. There are various systematic components of male sexual behaviour –

1. Sexual arousal
2. Courtship
3. Penile Erection and protrusion
4. Mounting
5. Intromission
6. Ejaculation
7. Dismounting
8. Refractoriness

- One component is the stimulator for the next consecutive one.

1. Sexual Arousal

- **Sexual arousal** – It is the 1st component of sexual stimulation representing the attraction between male and female. visual, auditory, and olfactory stimuli are mainly responsible for sexual arousal. Not only homosexual behaviour (cow and some pig breeds) but also estrous female secretes some pheromones to attract the male.

2. Courtship

- It is the activity performed by the male before the mating. It stimulates and prepares the female for copulation.
  - Sniffing and licking –

    female genitalia, urine and perineal region - Bull, ram & buck

    female head – Boar & stallion
  - Vocalization –

    Courting bleat – Ram & Buck

    Courting grunt – Swine

    Neighing – Stallion

    No vocalization – Bull
  - Nudging – (Gentle pushing to draw attention of female)

    With foreleg – ram & buck

    Nosing the flank – swine

    Biting over back or neck – stallion

- **Flehmen's reaction** – It is the sexual behaviour shown by the male during courtship. When a males sniffs the genitalia & urine of estrous female, because of the effect of pheromones on vomeronasal organ, it stands rigidly with extended head & neck and upcurling of the upper lip. It is shown by all species except boar.
- When buck enters into courtship behaviour, he shows kicking, pawing, and nuzzling the female, grunting, and displaying the flehman response (curling of the upper lip).
- The character of urine marking (marking the place of urinated estrous female) is of stallion. The Urinating pattern is not seen in cattle/buffalo bull and ram.
- Stallion display a characteristic precopulatory sequence beginning with nose-to-nose contact associated with a characteristic soft nicker vocalization. Sniffing, nibbling, nuzzling, and licking follow further interaction starting at the shoulder and proceeding to the abdomen & udder and then to the hind legs and perineum.
- The Arabian camel exhibits some signs when comes in contact with estrous female as extrusion of soft palate, gurgling sound, urine marking/ spraying, smudging the poll gland secretion and teeth grinding.
- Dog shows multiple unsuccessful mountings over estrous female during courtship.
- A normal male cat (selected by estrous queen) approaches with female, touches her nose with his nose, and then her perineal region. He may exhibit a flehmen reaction, by opening his mouth and closing his eyes while sniffing, an action associated with the reception of pheromones of the female by the male's vomeronasal organ. Tom approachs the queen from side or from back and grasps her neck in his mouth (not aggressively) and then positions himself to align with genitalia before intromission. Males grip estrous females an average of 15-20 seconds after being exposed to them, and achieve first intromission after an average of 1.5 to 2 minutes.

3. Penile Erection and Protrusion

- Penile erection is controlled by the centres situated in the cerebral cortex and lumbar spinal region. The erection of penis is under the control of parasympathetic nervous system.
- Contraction of ischiocavernosus muscles (penis erectile muscles) at the root of penis, and contraction of bulbospongiosus muscle cause increased

penile arterial blood flow and reduce the venous drainage leadsing to an increase the hydrostatic pressure in the penis which causes erection of the penis. Penis is protruded out of the prepuce due to penile erection along with relaxation of retractor penis muscles.

- There is to & fro penile movement of the erected penis before mounting. At the same time there is some fluid from the accessory sex gland (mainly Cowper's gland) comes out called "**dribbling**" in case of bulls. Dribbling is meant for cleansing the urethra (cellular debris and urine traces) in order to clear the passage for sperms.
- Just before mating, the bull rests his chin on female back when she is in standing estrous.
- In stallion, there is penile oscillation first and then penile engorgement (blood filling) takes place due to which penile rigidness is achieved.
- In tom, stimulation of parasympathetic nerves of the second sacral nerve roots produces an erection. Subsequent stimulation of the sympathetic nerves at lumbar 1-2 or of the hypogastric (sympathetic) nerve causes secretion of seminal fluid into the urethra and stimulation of the internal pudendal (i.e., parasympathetic, sympathetic, somatic) nerve results in ejaculation.

4. Mounting

- When more than one female is receptive, the bull will often preferentially mount the one that most recently comes into estrus. The bull may exhibit one or two unsuccessful mountings.
- The Male grasps female from behind firmly with forelegs during mounting. There is rhythmic pelvic swaying to seek the vulva before intromission.
- When there is mounting on the cat, Tom holds/grasps her neck with teeth.
- Camel mates with female in sternal recumbency. He makes the inexperienced female recumbent by pressing on her neck or back. He squats on the female and pushes himself toward the perineum region and directs the penis for seeking vulva in a partial erected condition.
- The dog continues mounting thrusting and dismounting till the successful intromission. During this activity, a small amount of prostatic clear sperm free fraction is ejaculate to clear the urethra from urine traces and any debris.

- Stallion and boar may also exhibit mounting and dismounting frequently before successful copulation.

5. Intromission

- There is a rhythmic movement of pelvic region and some contraction of abdominal muscles to seek the vulva. Superficial nerve endings of the glans penis are sensitized to detect proper heat and moisture at vulvar region.
- After Seeking the vulva, penis aligns the glans against the vulva and then intromission takes place with a pelvic thrust.
- Copulation time is least in cattle, sheep and goat (few seconds) than stallion (av 40 seconds) and boar (3 to 20 minutes). In camel it is about 3 to 25 minutes (average 5.5 minutes).
- Penile insertion and thrusting occur over a period of 20 to 25 seconds before ejaculation is observed in stallion.
- Intromission and adjustment of tom penis are helped by the paddling movement of the hind limbs by the queen.
- Full intromission is achieved after ejaculatory thrust in bulls. In camel, penis becomes fully extended after complete intromission.

6. Ejaculation

- It is under the econtrol of sympathetic nervous system. Abdominal muscle (rectus abdominis) contractions & nerve stimulation facilitate the ejaculation.
- Seminal ejaculation is coordinated by the lower lumbar and sacral region of spine along with medial optic area of the hypothalamus. Dopamine facilitates and serotonin inhibits the central control of ejaculation.
- Temperature and tactile sensation are the main stimuli in bull and ram & pressure is the main stimulus in stallion, boar and dog for intromission and ejaculation.
- Due to oxytocin release, semen transports through epididymis and vas deferens and contraction of sex glands force seminal content into urethra at the time of ejculation. Rhythmic contraction of urethral muscles, ischiocavernosus and bulbourethral muscles, semen is released out of the urethra.

- At the time of ejaculatory thrust (single thrust in ruminants), both the hind limbs are in the air and then the bull rests or presses his head over females back.
- Multiple pelvic thrust are seen in boar and dog. The head moves backwards at the time of ejaculation in ram & buck. Motionless male with scrotal contractions after ejaculation is the character of boar. Biting on the female his neck is seen during ejaculation in stallion. Camel stretches his body and extends neck during ejaculation. Ejaculation occurs several times during copulation in camel.
- In stallion, ejaculation occurs after 7 to 9 pelvic thrusts and there are 5 to 10 successive forceful ejaculations produced due to rhythmic contractions of the ischiocavernosus, bulbospongiosus, and urethralis muscles. Each ejaculation is evident by **tail flagging** (downward movement of the tail) and rhythemic contractions of hind leg muscles. Urethral peristalsis (palpable at the lower surface of penis) during ejaculation can be palpated. The time taken from intromission to ejaculate is about 30-45 seconds.
- In dog, copulatory tie or **coital lock** is seen due to swelling of bulbus glandis and constriction of vulvar muscles
- Queen exhibits painful crying at the time of copulation called as "**cat cry**" and after ejaculation queen stike out tom called "**rage reaction**". The queen usually does not allow sexual receptivity for 10-20 minutes after copulation. But after this time period she again allows male to copulate. Several matings are performed by tom within 30 to 60 minutes (average 7 intromissions) to ensure sufficient LH surge to cause ovulation as the cat is induced ovulator. Queen rolls and licks the vulva after copulation.
- Semen is deposited near the os cervix (bull, ram & buck) or in uterus (swine and stallion).
- **Pederasty** is an act of rectal intromission/copulation seen in boar. **Balling up** is masturbation by inserting penis into the prepucial diverticulum and **Onanism** (abortive ejaculate) is also a character in boar.

7. Dismounting

- Dismounting is more lethargic in buffalo bulls than cattle bulls. Penis is withdrawn back into prepuce after dismounting.
- After dismounting

  - Licking of the penis – Buck
  - Stretching the head and neck – Buck
  - No post coital reaction - Bull
- Stallion dismounts approximately 20 to 30 seconds after ejaculation and usually remains in proximity to the female for a few minutes. During this time, the stallion may engage in the investigation of any vaginal fluid and may display flehmen reaction.
- After dismounting in the dog, there is a face away (tail-to-tail) condition in which the penis is twisted through 180^0. There is occlusion of efferent penile veins which leads to slow veinous outflow i.e. very slow reduction in hydrostatic pressure. It maintains the penile turgidity for a longer period. At this period, 3rd fraction of ejaculate (about 30 ml sperm free accessory fluid) is pumped in to the vagina to flush the sperm rich fraction into uterus. After reduced penile blood occlusion along with vulvar muscle relaxation, the penile erection subsides and unlocking occurs. Coital tie may be about for 20 minutes (5 to 60 min).

8. Refractoriness

- It is the state of sexual exhaustion after ejaculation. It is the period between 1st ejaculate to second mount. During the period of refractoriness male shows no sexual interest in female after copulation.
- Buck, boar and stallion reach sexual exhaustion earlier than bull & ram i.e. more number of ejaculates can be achieved from bull & ram (bull can give 30-35 services/day which is the maximum).
- When a sexually exhausted male is exposed to a new female, he may restore his sexual activity sooner. This effect is called as **'Coolidge effect'**.

7

Infertility in Male Animals

Priya Ranjan Kumar and Vikas Sachan

Fertility is the ability of a male to impregnate the fertile female. The reduced or temporary loss of fertility is called infertility whereas permanent loss of fertility is called sterility. A male with low libido or poor sex drive may have excellent semen quality or a male with good libido may have poor seminal quality. So, libido has no direct relation with the male fertility. Male infertility can be categorized into Impotentia coeundi and Impotentia generandi.

1. Impotentia Coeundi

- The condition of reduced or lack of sex drive or Inability to copulate (failure of normal service) is referred to as impotentia coeundi.
- It may occur due to immaturity, poor experience to perform coitus, inability or unwillingness to mount, inability to perform intromission or failure of ejaculation.
- The following conditions may reflect the **immaturity or poor/bad experience** of the male animal which leads to impotentia coeundi:
  - Mating at a prepubertal or very young age
  - Rearing in isolation or a small confined room (poor learning to copulate)
  - Slippery floor, low roof, and unpleasant environment
  - Abusive handling and unfriendly attendant
  - Punishing or painful experience during mating
  - Continuous exposure to an estrous female
  - Furious and improper size of dummy
  - Presence of old and senior/dominant male at the mating place
  - Experience of too cold and too hot artificial vagina

- The following factors reduce the **sex desire** (poor libido) which leads to impotentia coeundi:
  - Deficiency of energy, protein, vitamins, TDN, minerals, etc.
  - Overfeeding (animals become obese and lethargic)
  - Obesity due to poor thyroid hormone (iodine deficiency)
  - Overfeeding of roughage (enlargement of the abdomen leads to lethargyness)
  - Over sexual use i.e. no sexual rest
  - Aged bull
  - Climatic and psychological factors
  - Poor exercise
  - Joint, nerve, and muscle pathologies
  - Endocrine imbalance (Poor testosterone to estrogen ratio)
- Libido may be improved by hormonal therapy such as hCG and GnRH as these hormones increase the level of testosterone but the effect is temporary. Overdosing or frequent administration of hCG may cause testicular edema resulting in abnormal spermatogenesis. Testosterone therapy (@ 100-500 mg in bull/stallion and @ 50-100 mg in ram/buck per week) may be used to increase libido temporarily but prolonged use may cause testicular degeneration and atrophy. Good management and exposure to estrous female may also be helpful. Dogs with poor libido may be trained with manual stimulation for ejaculation at regular intervals.
- Following painful or abnormal conditions which make male animals **unable and unwilling to mount**:
  - A pathological condition of foot/limbs, especially of hind limbs like overgrown hooves, interdigital growths, fibroma, sole penetration, etc.
  - Coxitis (hip joint inflammation) in dog, gonitis (stifle joint inflammation) in bull, tendonitis, and tarsitis etc.
  - Abnormal limb confirmation as sickle hock and camped behind.
  - A lesion in lower back trunk e.g. Honeymoon back.
  - Back pain (due to degeneration of the backbone, Ca deficiency,

pathology of vertebrae, or ageing). Animals may mount and dismount over and again indicating back pain during service test. Back pain is more pronounced in poorly exercised bulls. Over Calcium feeding leads to exostosis of vertebrae that may cause degeneration of intervertebral joints, stiffness, paraplegia, or increased tendency of fracture.

- Abnormal gait (spastic paresis, straddle gait, congenital ataxia, crampiness, etc.)
- Abnormal limb confirmation and joint angles: Abnormal limb confirmation and joint angles, especially of hind limbs may lead to impotentia coeundi. The weight of the whole bull is managed on the hind limbs during mounting and they are in full extension during the ejaculatory thrust. So, along with the **neuro-muscular system**, limb confirmation especially of hind limbs should be perfect for normal copulation.

Limb confirmation: To know the normal limb confirmation, observe the animal from the side and back. The angle of tarsal joint and metatarsophalangeal joint, when observed from the side, should not be too large or too small i.e. neither too curved nor too straight when observed from the side. A line/axis should bisect the stifle, tarsal and fetlock joint into two equal sagittal planes when observed from the back. Some abnormal limb confirmations:

- ***Sickle hock*:** The angle of the tarsus is too small and it may lead to lameness. Sickle-hocked stallion is prone to bone spavins that may lead to permanent lameness.
- ***Postiness/post-legged*:** The legs become too straight due to the large angle of the tarsus and the metatarsophalangeal joint becomes small. In young animal this condition is painless but as age advances, there is lameness with pain. This condition may pass on to progeny.
- ***Camped behind*:** The angle of the tarsus is too large as rear foot rests far behind. Such bulls sway their back frequently to find a comfortable position.
- ***Camped under*:** Animal keeps their hind legs underneath the body to reduce the pain and have a short stride. This condition is generally seen in horse.
- ***Bow-legged/narrow base/Toe-in*:** – Hind limbs are curved medially (hock joint moves outward) when observed from the back. The outside

toe may be curled and the outside wall of hoof is compressed. There may be lameness.

- ***Cow hock/wide base/toe-out/splay foot*:** Hind limbs are curved laterally (hock joint moves inside) when observed from the back. Inside wall of the hoof is compressed. There may be lameness.

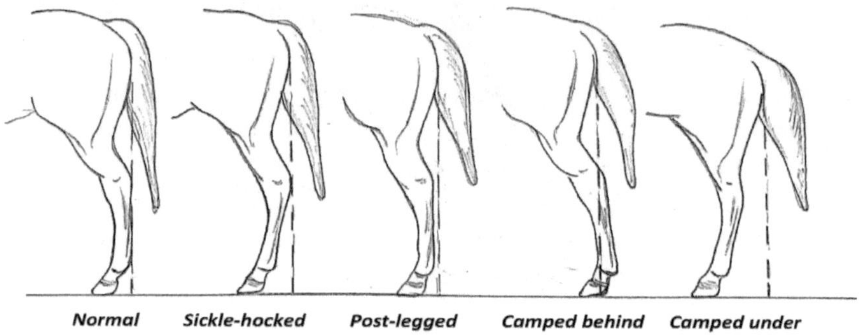

Rear leg confirmation from side

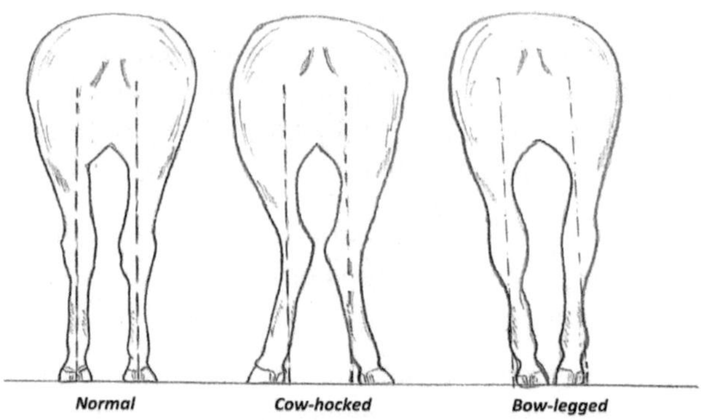

Rear leg confirmation from back

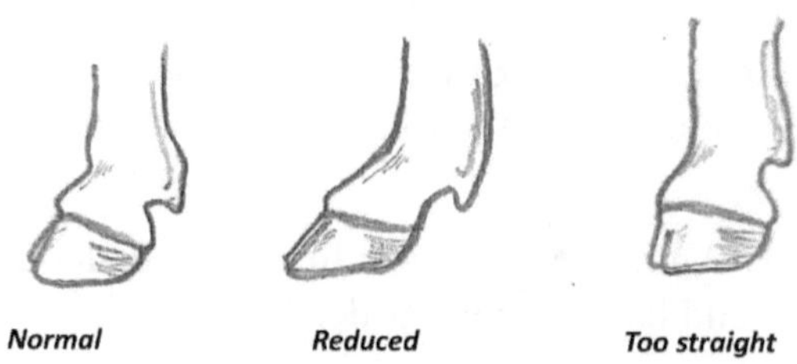

Confirmation of Metatarsophangeal joint

Neuromuscular abnormalities

- ***Progressive posterior paresis/crampy/neuromuscular spasticity/ spastic syndrome*:** It is characterized by spasm or inability to coordinate voluntary movement, usually seen in older bulls. It may occur due to abnormal motor function because of degenerating motor nerve fibres or spinal cord neurons. Due course of time, neuromuscular control at the anal and perineal region becomes abnormal and thus defecation and urination are not controlled. Intermittent spastic contractions of muscles of the back and rear limbs are also seen. When an animal rises or moves suddenly, there is a slight muscle spasm as in earlier signs but with progressing condition muscles of one or both hind limbs show spasmodic contractions for seconds to minutes.
- ***Spastic paresis (elso heel)***: It is characterized by full extension of hock and stifle joints due to chronic contractions of the gastrocnemius and superficial digital flexor muscles. This condition generally occurs in young bulls and affects unilateral limb muscles. Shaking/tremors in muscles are seen when the bull is rising.
- ***Peroneal nerve/Tibial nerve paralysis*:** Due to trauma, there is flexion of the hock and knuckling of the fetlock joint.
- **Honeymoon back:** It is characterized by the rupture of lumbo-dorsal fascia in some young and inexperienced bull due to quick and aggressive mounting. The bull experiences pain in lifting forelimbs (painful mounting) due to subsequent crepitation and swelling of muscles in the lumbo-dorsal region. Thus, the animal becomes reluctant to mount.
- **Failure of penile of intromission:** Some animals may mount but not be able to achieve intromission. It may be due to failure of penile erection or failure of penile protrusion.

Failure of penile erection may occur due to

- Abnormal parasympathetic nervous system
- Improper blood flow in the corpus cavernosum penis (CCP)
- Leakage of blood from CCP
- Presence of ectopic veins
- Rupture/broken penis
- Fracture of os penis (dysuria and haematuria)

- Atrophy of retractor penis muscles
- Preputial hair ring and biting abscess in tom

Failure of penile protrusion may occur due to:

- Short or under-grown penis
- Abnormal sigmoid flexure or retractor penis muscle
- Balanitis, posthitis, balanoposthitis
- Phimosis, paraphimosis, priapism
- Diphallus, phallocampsis
- Adhesion, necrosis, neoplasia in penis or prepuce
- Hypospadias
- Penile necrosis (strangulation due to preputial hairs or wool)
- Urolithiasis
- Exogenous drug administration (testosterone, glucocorticoids, anti-androgens, ketoconazole, cimetidine, anabolic steroids, etc) in tom
- Protruding intervertebral disc and degenerative myelopathy in tom

- Some psychological factors like an unknown or aggressive partner, painful experiences, stressful environment/surroundings etc. may also be responsible for the refusal of sexual arousal or mating process.

Ruptured/Broken/Fractured penis – It is the rupture of tunica albuginea due to abnormal pressure in CCP. The site of rupture is at the insertion point of retractor penis muscles or the dorsal side of the sigmoid flexure. It is most common in bulls. Bull shows arched back, stiffness, pain, and swelling with abnormal stepping etc.

- Generally occurs in young bulls with poor experience and strong sex urge
- Due to strong thrust with erected penis on the hind region of the cow elsewhere the vagina
- Hematoma, preputial oedema, or eversion of the prepuce may occur with normal urination.
- Soft hematoma may convert into adhesion, abscess, and hard fibrosis.
- Damage of penile nerves may also lead to failure of penile erection and intromission.

- It may be treated by evacuation of hematoma by surgery, suturing of T. albuginea, sexual rest for 2-3 months, antibiosis, anti-inflammatory drugs, and massaging.
- Penile trauma or trauma to os penis in the dog may prevent mating.

Phallocampsis (penile deviation) – "Lateral or ventral deviation of the penis"

- Lateral deviation of the penis is due to a congenital defect of tunica albuginea or poor development of the dorsal apical ligament. Adhesion after trauma to the penis or prepuce may also cause deviation of the penis.
- Ventral deviation of the penis (**rainbow penis**) is due to persistent penile frenulum (non separation of penis and prepuce at puberty), defect in the tunica albuginea, congenital defect or adhesion and fibrosis of penile tissues, etc. Normally persistency of the frenulum beyond 11 months of age is considered as abnormal. The penile frenulum may persist due to rearing the bull in isolation i.e. no exposure of riding behaviour.
- The persistent frenulum may be transacted surgically after ligation of the blood vessels to subside the condition of ventral penile deviation.
- The dorsal penile apical ligament is fixed with tunica albuginea with the help of fascia lata (taken from biceps femoris muscles) to prevent slipping or misplacing of the ligament to correct the penile deviation.
- **Corkscrew penis** – Normally bull penis becomes spiral after ejaculation but in some bulls, spirallng occurs prematurely within prepuce or just after touching the cow/dummy perineal region (premature spiral deviation of the penis; **PSDPs**). It may be due to abnormal dorsal apical penile ligament (inherited or acquired) or some neural pathology.

Balanoposthitis –"Inflammation of the penis and prepuce".

- Infection and inflammation of the penis is called **balanitis** while Infection and inflammation of the prepuce is called **posthitis**
- There is pain, adhesion, and stenosis of the preputial opening. Laceration and ulceration on prepuce with dermatitis may also be found
- Condition is common in bull, dog, and ram but less common in stallion, boar, and tom.
- Common organisms causing balanoposthitis are E. coli, streptococcus sp, staphylococcus sp, C. pyogens, M. tuberculosis, aspergillus, mucor, mycoplasma, IBR-IPV virus, and calicivirus, etc. Ureaplasma infection causes granulomata formation on the penis.

- E. coli, canine herpes virus, calicivirus, etc may cause balanoposthitis in canines. **Orf** (contagious pustular dermatitis/ecthyma contagiosum/ scabby mouth) may cause balanoposthitis in ram & buck. BoHV-1 (bovine herpes virus) may cause ulceration and purulent balanoposthitis in bulls causing painful urination and intromission. Dourine (by Trypanosoma equiperdum), habronemiasis (by larval infection of Habronema musca), and equine coital exanthema/horse pox (by equine herpes virus EHV-3) cause balanoposthitis in equines.

Pizzle rot/ Sheath rot -

- It is ulcerative balanoposthitis in ram. Its main causative agent is Corynebacterium renale.
- A high protein diet causes ammonia scalds at the prepucial region (High ammonia is produced from urea by the action of bacterial urease enzyme). Trauma due to faulty positioning of the artificial vagina (AV) or hair drawn in prepuce during semen collection may also cause balanoposthitis.
- Ulceration, necrosis, and scab/crust formation on prepucial mucosa may cause stenosis and blockage of the preputial opening.
- It can be treated with the use of broad spectrum antibiotics (penicillin & cephalosporins), anti-inflammatory drugs, antiseptic douching of the prepuce, washing with $KMnO_4$ water or with 1% H_2O_2 solution, bacitracin ointment application and sexual rest.

Phimosis – The inability of the male animal to protrude penis out of prepuce due to a stenosed preputial orifice (congenital or due to injury) is called phimosis.

- It may be also due to a small penis, penile neoplasia, balanoposthitis, or prepucial adhesion. It is more common in dog and stallion but also be recorded in other domestic animals. Premature swelling of penile erectile tissues before protrusion in dogs due to overexcitement may also make the penis unable to protrude out.
- If the condition is congenital, the preputial orifice can be enlarged by surgical removal of a small piece of mucosa from the ventral side of the prepuce. If due to injury, antibiosis and anti-inflammatory drugs may be taken care.

Paraphimosis –The inability of the male animal to retract back the penis in preputial sheath after protrusion is called paraphimosis. It may be due to abnormal or damaged retractor penis muscle, swelling/inflammation of the penis &prepuce, local penile injury, stenosis of the preputial orifice, balanoposthitis, penile neoplasia, damaged pudendal nerve or penile paralysis due to spinal injury. During prolonged protrusion of the penis and prepuce, venous drainage may be impaired due to the preputial orifice ring which aggravates the condition. It is more common in dog and stallion.

- It can be treated by applying cold packs, lubrication, antiseptic ointments, diuretics, and replacing the penis in prepuce after compression gauzing or/and incising the prepucial orifice.
- Phallopexy (fixing of the penis in prepucial sheath) may be performed to keep the penis at its anatomical place till the condition subsides. Antibiotic and anti-inflammatory drugs are administered.
- If the condition is more severe or incurable then penectomy (surgical removal/amputation of the penis) may be performed.

Priapism – It is persistent penile erection without sexual stimulation for more than 4 hours. It may be due to increased arterial blood flow (nonischemic or high flow priapism), reduced/occluded venous blood flow (ischaemic or low flow priapism) in the penis or abnormal neural control of erection. It is more common in dog, stallion, and tom.

- Urine retention and painful urination may be there. In chronic cases, oedema and necrosis of the penis may occur.
- Cold pack application, lubricants, antibiotics, anti-inflammatories, anticholinergics and diuretics may be beneficial. Intravenous injections of benzatropin mesylate and adrenaline into CCP may be beneficial up to some extent. Surgical drainage of CCP blood may subside the condition, but the ultimate remedy is penectomy.

Diphallus – The presence of a double penis is referred to as diphallus. It is a congenital anomaly. Penis may be completely or partially duplicated with variable symmetries. Diphallus prevents normal copulation.

Hypospadias – It is a congenital anomaly of abnormal extrapelvic urethral formation. Urethra has an anabnormal opening (at the ventral aspect of the penis) site due to abnormal penile development.

- Penile/balanitic hypospadias – the urethra opens ventrally caudal to the penis and cranial to the scrotum.
- Scrotal hypospadias – The urethra opens at the scrotum
- Perineal hypospadias – The urethra opens at the perineal region.

Epispadias – When the urethra has an opening at the dorsal aspect of the penis.

Preutial eversion: Preputial mucosa is protruded/everted out of the sheath permanently which subjects to inflammation. It may be due to aplasia or hypoplasia of the retractor muscle of the prepuce.

Penile neoplasm

- Fibropapilloma in bovines (due to bovine papilloma virus)
- Squamous cell carcinoma / squamous papilloma and in stallion
- Transmissible venereal tumour in the dog (venereal granuloma/ lymphosarcoma/sticker tumour)
- They can be treated with topical application of fluorocil (in cases of small tumours), anticancerous drugs, surgical removal of tumours (laser and cryosurgery), and penectomy.

Intersex

- True hermaphrodism (presence of both testes & ovarian tissue along with both types of externalia),
- male pseudohermaphrodism (Tetses with external clitoris),
- Female pseudohermaphrodism (Ovary with male external genitalia). These are some more conditions leading to infertility/sterility in male animals.
- Some other conditions like preputial prolapse, fracture of os penis and calcification of retractor penis muscle, etc also lead to failure of intromission i.e. impotentia coeundi.
- Hemospermia/Hematospermia (blood in semen due to trauma, inflammation, CSP leakage, etc) may also cause infertility. It is common in stallion & bull.
- **Failure of ejaculation:** Some male animals mount and intromission successfully but fail to ejaculate. It may be due to -
  - Congenital absence or abnormal penile receptors on glans
  - Damaged sensory dorsal nerve of the penis
  - Compression of spinal nerves
  - Poor sympathetic nervous system

- Localized back pain
- Peritonitis in caudal abdomen

2. Impotentia Generandi

It is characterized by failure of fertilization/conception even after normal copulation. It may be due to disease of the testes, epididymis, and accessory sex glands where semen may be abnormal (abnormal seminal attributes such as sperm morphology, concentration, motility, etc). It may be due to apparently normal semen with sperm defects (inherited sperm defects).

A. ***Impotentia generandi with abnormal semen*** – It occurs in the following conditions:

- Testicular hypoplasia
- Testicular degeneration
- Orchitis
- Cryptorchidism
- Seminal vesiculitis
- Prostatitis
- Ampullitis

Testicular hypoplasia – It is characterized by the incomplete development of germinal epithelium of the seminiferous tubules due to reduced germinal cells in the testes. This lack of germ cells (spermatogonia) may be due to:

1. Failure of germ cells to develop in the yolk sac
2. Failure of germ cells migration to the genital ridge
3. Failure of germ cells to multiply in gonad
4. Degeneration of embryonic germ cells

- Affected testes become small (reduced scrotal circumference < 30 cm in bull), flabby and firm with a shorter spermatic cord. Testes are freely movable in the scrotal pouch unlike to testicular atrophy condition. It may be bilateral or unilateral (more common in left testes). Animals are sterile if it is bilateral testicular hypoplasia.It may be congenital and hereditary (controlled by recessive autosomal gene). It is common in ram, bull, and stallion. Klinefelter's syndrome is a cause of testicular hypoplasia in bulls, ram, boar & dogs.

- Spermiogram in testicular hypoplasia:
  - No or fewer sperms (aspermia in bilateral & oligospermia in unilateral condition)
  - Low motility
  - Abnormal morphology of sperm
  - More sperms with proximal protoplasmic droplet
- In testicular hypoplasia, gonadal sperm reserve is poor due to reduced/ no spermatogenesis i.e. semen may be oligospermic or aspermic. There is no effect on leydig cells i.e. normal libido.
- **Giant cells** (multinucleate cells; incomplete maturational division i.e. nuclei divide but not cytoplasm) and **medusa cells** (ciliated cells from efferent tubules) are commonly observed in testicular hypoplasia.
- Condition is hereditary and poorly diagnosed till two years of age in bull or well grown other small animals. There is no treatment for testicular hypoplasia.

Testicular degeneration – It is the degeneration of the germinal layer of seminiferous epithelium including spermatogonia and Sertoli cells impairing spermatogenesis. It is the most common cause of male infertility.

- Initially in acute cases, because of inflammation, heat & pain, the testes may be enlarged and tense. Cauda epididymis is softer and smaller than normal. Due course of time, the seminiferous epithelium is atrophied and testicles become small & softer. Later on, due to calcification & fibrosis testicles may become firm and hard and least resilient.
- Germinal epithelium is most susceptible, while leydig cells are found to be most resistant to degeneration. The animal becomes sterile due to degeneration of germinal epithelium, but libido/sex drive may be maintained (as leydig cells are least affected).
- Testicular degeneration may be reversible or irreversible according to the severity of the lesion. The rate of degeneration is very fast (hours to days) but regeneration is very slow (weeks to months). Regeneration depends on the degree of damage to seminiferous tubules. In germinal epithelium, sertoli cells & spermatogonial cells are most resistant to degeneration i.e. once they are found degenerated along with a blocked lumen in the affected testicle, regeneration will not be there.

- Degenerated testicular tissue may result in the production of anti-sperm antibodies that might adversely affect the sperms produced by the normal testis.
- It may occur unilaterally, if the lesion is localized and bilaterally, if the lesion is generalized.
- Spermiogram in testicular degeneration condition:
  - Normal seminal volume
  - More immature round spermatogenic cells
  - Reduced sperm motility
  - Low sperm concentration (oligozoospermia or azoospermia)
  - Increased dead sperms (necrozoospermia)
  - More abnormal sperms (more PCD & detached head etc)
  - Diadem defect may be seen temporarily
  - Presence of giant (multinucleate cells)
- The lower level of plasma estrogens with high plasma FSH and LH is an indicator of low fertility and possibly testicular degeneration in stallion.
- The duration between testicular degeneration and the appearance of abnormal spermiogram with defected spermatozoa

  Bull - 60-70 days

  Ram - 60-70 days

  Stallion - 40-45 days

  Boar - 50-60 days

  Dog - 60-70 days

Causes of Testicular Degeneration

1. Temperature

- High environmental temperature (summer & humidity) or Direct heating
- Unshorn or excessive wool over the scrotum of ram
- Frostbite (inflammation & necrosis)
- Excess fat deposition over the scrotum (no heat loss) due to high protein and high energy diet.

- Scrotal mange (by Chorioptesbovis) in ram
- Local irritant, Inflammation, scrotal dermatitis, ectopic testes
- Inguinal/scrotal hernia (disturbed thermoregulation)
- Testicular lesion/trauma

2. Irradiation – (X-rays, Gamma rays, etc.)

- Most susceptible – spermatocyte
- Most resistant – leydig cells &sertoli cells
- Sperm concentration reduced with increased abnormality

3. Hormonal imbalance

- Sertoli cell tumour (common in dog) – excess estrogen production
- Leydig cell tumour – excess testosterone production
- Adrenal tumour – excess corticosteroid production.
- Tumours of the pituitary &hypothalamus –abnormal GnRH & gonadotrophin
- Excess estrogen or testosterone leads to negative feedback effects on the hypothalamus and pituitary i.e. reduced GnRH and gonadotropins. Similarly, corticosteroids (stress) also hamper the production of gonadotrophins (LH). It causes testicular degeneration.
- Testicular neoplasia, atrophy and degeneration along with obesity in dogs is called dystrophia adipogenitalis.

4. Age

- Overaged males are prone to testicular degeneration (progressive and permanent) due to poor management, and disease conditions.
- Testicular degeneration is common in dogs and bulls at >10 years of age and in tom > 12 years of age.

5. Stress/Trauma/Disease –

- Shipping, heat, cold, severe fatigue, improper housing
- Abscess, contusion, arthritis, laminitis, Myiasis, and foot rot etc
- Pneumonia, peritonitis, IBR, JD, FMD

- Stress leads to increased corticosteroid secretion and reduced gonadotrophins (e.g. LH) secretion which may cause testicular degeneration.

6. Systemic/localized diseases

- Localized disease of testes due to pressure, heating, inflammation
- Inflammation causes heat, oedema and congestion
- Orchitis
- Trauma and abscess of testes & scrotum
- Brucella, mycobacterium tuberculosis, Corynebacterium pyogens, etc
- IBR-IPV, BEV (Bovine enterovirus), epivag, and canine distemper infection cause orchitis, testicular degeneration, and testicular fibrosis.
- Obstruction of the efferent duct may cause testicular degeneration because of continuous back pressure.

7. Nutrition

- Under-feeding and poor nutrition leads to debilitating diseases and stress.
- Under-nutrition causes decreased gonadotrophins.
- Vit. A and Zn deficiency may also cause testicular degeneration.
- Deficiency of vitamins A, B, C, and E, amino acid, and minerals may aggravate the condition.

8. Poisons

- Dipping of ram in arsenical solution or other highly concentrated ectoparasiticide solutions
- Use of antimony compound in dogs for treating heartworm disease (Dirofilariaimmitis)

9. Autoimmunization

- Faulty or experimental injection of autologous testicular material
- Damage of the blood testes barrier due to any trauma

10. Testicular Tumours

- Tumours of interstitial cells, sertoli cells, germinal epithelium, etc.
- Excess steroid production due to tumuor causes testicular degeneration Common in old age dog

11. Vascular Disturbance

- Abnormal blood circulation in testes e.g. Varicocele of the spermatic vein (abnormal heat exchange).
- Strongyle larvae in a testicular artery in stallion may cause vascular disturbances in the testes. Vasculitis, neoplasia, granuloma formation, torsion of the spermatic cord, etc. may also be responsible.
    - There are other several conditions responsible for testicular degeneration like testicular fibrosis, testicular calcification, epididymis tumour, aplasia of the mesonephric duct, etc.
    - Diagnosis of testicular degeneration can be done with testicular biopsy, testicular histology (presence of giant cells, cytoplasmic vacuoles, thinning of germinal epithelium, presence of lymphocytes & plasma cells in seminiferous tubules, fibrosis, etc), and trans-scrotal ultrasonography. History of infertility, testicular trauma, and seminal picture may also indicate the condition.
    - The prognosis depends on cause and severity. On biopsy, if the seminiferous tubule is found patent and has an intact basement membrane with spermatogonia then the prognosis is good. Chronic cases with abscess and bilateral tumours etc make the prognosis poor.

Treatment is according to the causative factor of testicular degeneration discussed above as

- Balance diet with minerals, vitamins, and good protein
- Supplementation of especially Vit A and Zn in the feed
- Proper management like cooling to avoid heat stress during the hot season
- Regular exercise, deworming and vaccination
- Sexual rest with antibiosis and anti-inflammatory in case of orchitis
- Removal of the testes in acute unilateral orchitis

- Removal of the testes in case of testicular tumour
- On removal of degenerated testis often results in hypertrophy of the remaining testis and a resultant increase in sperm numbers.
- Treatment of local and systemic diseases
- Hormonal therapies (testosterone, FSH, Thyroxine) can be done but they are of little value.
- Spermatozoa with less morphological abnormality and increased motility indicate recovery. Testicular degeneration doesn't progress once the cause is removed but even after treatment, generally seminal quality is not regained fully as previously.

Orchitis – "Infection & inflammation of testes"

- Continuous inflammation may lead to suppuration (pus formation and accumulation) and then necrotic destruction of testes. Unilateral orchitis is more common.
- It may be caused due to
  - Trauma/wound
  - Bovine enterovirus in bulls
  - Brucella abortus, B. canis, B. suis, B. melitensis
  - Tuberculosis in bulls may cause granulomatous orchitis
  - Mycoplasma, Aranobacteriumpyogens, Histophilussomni
  - Ascending infection from urethra
  - Descending infection from accessory sex glands
  - Canine distemper and mycoplasma in dog
  - Autoimmune destruction of testes in dog
  - Damage to blood-testes barrier
  - Feline infectious peritonitis virus and epivag virus infection
- In acute conditions, there is hyperaemia, swelling (2-3 times enlarged), and pain with hotness but as chronicity increases, testes adhere to the scrotum and become shrunken, and fibrotic with leaking abscess. Semen is loaded with white blood cells.
- It can be treated with

    - Broad spectrum antibiotics and anti-inflammatory drugs
    - Application of antiseptics and ice packs
    - Glucocorticoids
    - Sexual rest
    - Removal of affected testes
    - Castration in case of bilateral orchitis
- Once treated, animal should be avoided for breeding purposes.

Cryptorchidism

- It is the condition of failure to descend one (unilateral) or both (bilateral) testes in the scrotum. When epididymis or a part of the testicle remains in the inguinal canal, it is called incomplete cryptorchidism. The condition may be inherited as sex limited autosomal recessive trait. It is more common in stallion, dog and boar.
- Cryptorchid testes are smaller, soft and flaccid than normal. Unilateral cryptorchid animal is usually fertile with increased (compensatory) size of scrotal testes. Bilateral cryptorchidism is characterized by oligospermia or aspermia which makes animal infertile/sterile.
- There is poor/no spermatogenesis (due to abnormal testicular thermoregulation) in the cryptorchid testes. Spermatogenesis may occur up to spermatogonia A or B in the cryptorchid testicle and up to the primary spermatocyte in the inguinal testicle. Leydig cells are unaffected i.e. libido remains normal.
- In foal, if at the time of birth, the gonad does not come into the scrotum, chances of cryptorchidism are more because after birth size of the testicles begins to increase and the vaginal ring constricts. The right testicle is smaller in foal and descends into the scrotum before the left one.
- Cryptorchidism may be because of defective gonadotrophin production, mechanical defect related to gubernaculum abnormality, genetic defect (inherited), testicular chromosomal abnormality, and poor testicular production of androgens. Higher total estrogen is produced by cryptorchid testes in stallion and boar than normal scrotal testes (because destroyed sertoli cells secrete no inhibin i.e. no inhibition of FSH) but the level of estrone sulfate is found to be very low in the absence of testicular tissues in the stallion.

- Cryptorchid testes are more prone to the development of neoplasia (seminoma &sertoli cell tumour in dogs and teratomas in stallion).
- A definitive diagnosis of monorchidism is based on history, behaviour, physical examination, hormonal testing, and probably surgical exploration. Ultrasonography, low basal testosterone level (e.g. <0.24 ng/ml in stallion) and measurement of estrone sulfate in stallion may be helpful in diagnosing the condition.
- The animals with testes in the inguinal canal or upper region are called high flankers.
- Some conditions like anorchia (No testes) and monorchia (only one testis) may also be found. Polyorchia (more than one supernumerary testis) is rare.

Testicular neoplasia

- Generally, tumours occur in cryptorchid testes. It may lead to torsion of the spermatic cord.
- Common in dogs (Interstitial cell tumours are more common than of sertoli cell tumours). There may be prostatic hyperplasia due to excess androgen production in the case of interstitial cell tumours. Seminomas may also develop in dogs.Testicular neoplasia is rare in stallion (seminoma & teratoma), boar, bull & ram.
- There is increased estrogen plasma concentration in the case of a sertoli cell tumour. Feminization, penile atrophy, gynaecomastia, etc are common sequels.
- Teratoma, seminoma and teratocarcinomas are germinal neoplasms (arise from germinal cells) in stallion. Teratoma is a benign tumour with hairs, teeth, bone, cartilage, etc within. Seminoma is a germ cell tumour (Malignant type) that develops in the germinal layer of seminiferous tubules. Sertoli cell and leydig cell tumours are nongerminal neoplasms. Most of them are benign tumours.
- Removal of tumours (cryosurgery etc.) or removal of neoplastic testes may be the remedy.

Seminal vesiculitis

- It is the infection and inflammation of the seminal vesicle. Seminal vesicle may be enlarged, firm, tense and painful. Semen has purulent exudates. In chronic condition, the gland is of reduced size, fibrosis without pain, and lost lobulations. Semen has a brownish colouration (degenerated blood from inflamed gland) with pus.

- The main causative agent for vesiculitis is Arcanobacterium pyogens (trueperella pyogens). Other agents are Brucella abortus, Chlamydophila, Epivag, IBR (Infectious bovine rhinotracheitis)/IPV (Infectious pustular vulvovaginitis), Mycoplasma bovis and ureaplasma diversum, etc. Corynebacterium renale, E. coli, Streptococci, Staphylococci etc aid in disease condition.
- Condition is common in bulls of younger age (1-3 years). It can be diagnosed with rectal palpation or transrectal ultrasonography.
- Sperms are with reduced motility. Semen has high pH, low fructose, and more polymorphonuclear leukocytes.
- In less severe cases antibiotics (macrolide), and anti-inflammatory drugs are helpful with sexual rest. In severe cases, the gland may be amputated or orchidectomy is recommended. Injections of ceftiofur or penicillin directly into the gland are more beneficial and may subside the condition in 3 weeks.

Prostatitis

- It is an infection & inflammation of the prostate gland (common in old age dogs). There may be hypertrophy and/or hyperplasia of the prostate gland. It may lead to diffused/local abscess formation in the gland, haematuria, dysuria, urethral bleeding with leukocytes in urine, etc
- Causative agents are bacterial (Brucella, E. coli, Streptococci, Staphylococci Sp. etc) and mycoplasma infection
- It may also occur due to high androgen because of interstitial cell tumours.
- It can be treated with progestogens. Castration and anti-androgenic drugs (e.g. finasteride) may be useful.

Ampullitis

- It is an infection and inflammation of the ampullae. Ampullae become enlarged, firm, and painful which may lead to blockage. Condition is more common in stallion.
- Spermiogram shows oligospermia with abnormal sperms (immotile & detached head sperms).

- Ampullary massage, oxytocin, and increasing ejaculation frequency may be the treatment.

Spermiostasis – No seminal flow due to blind rudimentary mesonephric duct or defective efferent duct.

Funiculitis – Infection and inflammation of the spermatic cord.

Epididymitis – "Inflammation of epididymis". It is common in rams (by Brucella ovis). Actinobacillus seminis and histophilus ovis may also cause epididymitis in rams. It may cause orchitis, epididymal obstruction granuloma formation, and poor seminal quality.

Sperm granuloma – There is sperm leakage due to rupture of the epididymis because of duct obstruction pressure (back pressure) or because of faulty vasectomy. Granuloma is formed as a result of antigenic inflammatory reaction because of these foreign antigens (leaked sperm) which accumulate in the surrounding stroma. Sperm granuloma is characterized by high content of lipid and mycolic acid produced by degenerated sperms. It is more commonly observed in bucks. Libido may be normal in this condition.

Testicular torsion (Torsion of the spermatic cord) – It causes obstructed blood flow and lymphatic. It is common in stallion with signs of colic and swollen painful testicles filled with scrotal fluid. Surgical correction, orchiopexy (to avoid recurrency) or castration is recommended.

• Testicular hematoma, testicular abscess, testicular cyst, etc are some more conditions affecting adversely the testes and hence the fertility.

B. *Impotentia generandi with apparently normal semen or abnormal sperms*

- In some cases, semen quality is apparently normal but sperm is not able to fertilize the ova or if able, early embryonic death will occur due to various sperm abnormalities.
- Seminal volume, colour, pH, and sperm motility looked normal but the sperms are infertile, which may be due to abnormal nuclear material or abnormal nuclear shape, abnormal sperm metabolism, and abnormal acrosome etc.
- Some common sperm abnormalities are:

Diadem defect

- There is an invagination or pouch formation in the nuclear material of the sperm cell.

- A beaded structure (pearl necklace) at the equatorial segment of the anterior border of post nuclear cap can be seen. Single large vacuoles (in pouching) may be there on th esperm head due to which head shape is distorted.
- This defect is due to abnormal spermatogenesis. This defect is categorized as primary, major and uncompensable defect.
- It can be diagnosed by Eosin-nigrosin stain, fuelgen stain/wet mount observing under phase contrast microscopy or differential interference microscopy
- Normal sperm motility and concentration but no/ abnormal fertilizing capacity is there.

Knobbed acrosome sperm defect

- Accentrically placed thickening of acrosome
- Defected spermiogenesis involving the golgi complex
- Inherited, autosomal, recessive and sex-linked sperm defect
- Common in bull, boar and dog
- Diagnosed by eosine-nigrosine, eosin-B, fast green stain observed under phase contrast microscopy
- Normal sperm motility and semen concentration
- Sperm is unable to penetrate ova
- Primary, major and uncompensable sperm defect
- A high percentage of sperm defects related to the head, midpiece, and tail due to abnormal spermatogenesis, maturation (protoplasmic droplets) etc also lead to infertility.
- Acrosomal crest, acrosomal fold, detached or abnormal acrosome of spermatozoa may also relate to reduced fertility. Sperm chromosomal aberrations, and abnormal nuclear protein formation during spermatogenesis result into infertile sperm formation. Inheritance of disturbed enzymatic activity hampers the normal sperm metabolism which leads to sperm death.

8

Coital Injuries in Male Animals

Vikas Sachan

- These are the injuries occurred to male animal at the time of natural mating. These injuries do not occur only to genital organs but also to other organs or muscles.
- Most coital injuries occur to male animal with good libido (high testosterone level) but having aggressiveness and poor experience of mating. Such animals tend to quick response in the presence of estrous female.

Some of the coital injuries are:

1. Balanitis and Posthitis

- Trauma to the penis and prepuce which later on leads to infection and inflammation.
- This occurs especially in young inexperienced bulls with great sexual desire
- Adhesion may lead to phimosis or paraphimosis condition
- IBR-IPV infection from cow genitalia after mating results in pus formation and necrosis of the penis and prepuce.
- Prepucial eversion leads to laceration and bleeding.

2. Penile Trauma

- Kicking by the dummy or female to the mounting male
- Improper alignment of penis toward vagina before thrust (Striking against perineum)
- Excessive bending in erected penis may lead to penile fracture (broken penis) generally at the point of attachment of retractor penis muscles which leads to hematoma. The penile fracture may occur at the time of

coital lock in the dog. There is a rupture of the corpus cavernosum and tunica albuginea.

3. Penile Hemorrhage

- Bleeding from penile tumour, polyp or ulcer during or after mating
- Urethral polyps or calculi
- Laceration of the penis or prepuce during coitus
- Penile fistula (up to corpus cavernosum)
- Little cut on the penile tip or prepuce during trimming of preputial hairs
- Penis may be traumatized and bleed due to improper alignment of the artificial vagina at the time of semen collection.

4. Testicular and Scrotal Injuries

- Kicking during coitus may lead to crushing or blow injury to the testicles.

5. Herpes Coital Exanthema

- Ulcerative lesions at the penis and prepuce due to equine herpes virus
- Transmission at the time of coitus

6. Other Injuries

- Musculo-skeleton injuries occur due to falling or slipping of the male who is quite smaller than the female. Falling and misbalancing is also due to aggressive and uncontrolled thrust at the time of coitus.
- Ventral hernia due to kicking by female during coitus
- Limb fractures, hip dislocation, gonitis, muscle or tendon rupture due to slipping
- Inguinal hernia with strangulation of the intestine may occur after service in stallion
- **Honeymoon back** i.e. rupture of lumbo-sacral fascia during uncontrolled and aggressive thrust by young bull
- Harm to tom by a furious strike by the cat during rage reaction.

Treatment is according to the cause and severity

- Antibiotics, local antiseptics and anti-inflammatory drugs are used accordingly.
- Surgical removal of tumourous growth, polyp or calculi in the urethra
- Surgical correction of ruptured tunica albuginea.
- Sexual rest provides sufficient time for healing and subsiding inflammation
- Proper training of the bull before mounting
- Use of a docile dummy or familiar female during semen collection or mating
- Proper use of AV at the time of semen collection.

9

Vices of Male Animals

Vikas Sachan and Jitendra Kumar Agrawal

- These are the bad habits of the male animal due to which ultimately their maximum reproductive potential cannot be utilized.
- There are so many vices seen in male animals like stall walking, weaving, cribbing, wind sucking, crib biting, viciousness etc. These behavioural disorders are mostly due to improper managemental practices like abusive handling, poor exercise, irregular semen collection and poor exposure to female etc. These may be also due to hormonal imbalance or nervous disorders.

- Some important vices which are more related to reproductive behaviour are:

1. Masturbation (Onanism)

- Bulls masturbate by rubbing their prepuce and penis against any object.
- Boar masturbate and ejaculate by inserting the penis into the preputial diverticulum called **Balling up**
- Boar may show rectal copulation and leads to ejaculate : **Pederasty**

Reasons

- Prolonged sexual rest or no female exposure for a long period
- Accumulation of smegma in prepucial sheath of stallion
- Poor exercise

Management

- Semen collection or mating/exposure with female at regular interval
- Regular cleaning of smegma at the preputial sheath with soap water or any cleansing agent

- Use of stallion shield or bird cage device over glans or use of an adhesive tape behind glans penis
- Regular exercise
- Regular washing and cleaning of the penis and preputial sheath
- Loose sutures with stainless steel at the preputial orifice in bull, ram and buck
- Surgical removal of preputial diverticulum in boar
- Castration in dog

2. Viciousness

- Aggressiveness and furiousness
- Difficult to control during management and breeding practices

Reasons

- Rearing in confinement
- Treatment, teasing, irritating, abusive handling and unnecessary punishments
- High steroid (testosterone) level

Management

- Proper training with gentle handling
- Regular exercise
- Castration

3. Slowness in Breeding

- Male is slow in approaching female for mating
- May be reluctant to mate

Reasons

- Poor libido (low level of testosterone)
- Painful experience due to any accident during previous copulations
- Improper, rough and harsh handling
- Poor experience and improper training

Management

- Proper training and regular exercise
- Use of docile female or dummy during the training period
- Gentle handling and proper management

10

Methods of Semen Collection

Vikas Sachan and Anuj Kumar

Objectives

- Seminal evaluation and processing
- Evaluation of bull fertility on the basis of its seminal characters
- To make the Artificial insemination programme successful

History Related to Semen Collection

- The use of artificial vagina was started by Milovanov for semen collection. Before that method of placing the cotton ball in vagina was used.
- AV for Dog was discovered by Amantea in 1914. Milovanov discovered AV for Bull, ram and stallion. Mckenzie discovered AV for boar (1931).
- The electroejaculatory method of semen collection was initiated in a bull by Laplaud & Cassou and in ram by Gunn.
- Burrows and Quinn (1937) developed the method of abdominal massage and pressure to collect semen in poultry.

Semen Collection in Bull

- The training of a breeding bull is usually started at the age of 18 months and semen collection can be done at 24 months of age
- There are 4 methods of semen collection in bulls –
  1. Directly from the vagina
  2. Rectal massage method
  3. Electro-ejaculatory method
  4. AV (Artificial vagina) method

1. ***Directly from the vagina***

- It is the simplest and oldest method. With the help of a spoon or long nozzle fitted syringe, semen is aspired from the vagina of a recently mated (natural mating) female.
- Seminal quality is not very good because of mixed vaginal mucus (estrual discharges)

2. ***Rectal massage method***

- In this method, there is no penile erection or ejaculatory thrust is required. Seminal vesicles and ampullae are massaged per rectally in a fashion of pressing / milking caudally in a restrained bull.
- Stimulation of bull with exposure to female helps in the collection of improved semen quality (volume and sperm concentration). Also, because of some straightening of sigmoid flexure, all the retained semen comes out i.e. more semen can be collected
- This method is useful for semen collection in crippled, weak, and unable to mount bulls.
- Semen is donated in the form of dribbling i.e. more chances to get contaminated with prepucial hairs, dirt, bacteria etc. Also sometimes due to the pressing of urinary bladder, urine can be mixed with semen which reduces the quality of semen. Obtained semen is good in volume but has poor sperm concentration

3. ***Electro-ejaculatory method***

- The Electro-ejaculatory method in bull was discovered by laplaud & Cassou in 1948.
- After proper restraining of the bull, rectal emptying and washing with 6% NaCl is done (for better electrical conductivity).
- An electrical probe (provided with an electric power pack) of length 35-60 cm and 4-8 cm diameter is used. The probe is lubricated and inserted into the rectum up to 30-40 cm. Keep gentle pressing on the ventral floor with rising AC current from 0V to 5 V and again decline to 0V within 5-10 seconds. This procedure can be repeated for 2-3 times at the interval of 5-10 seconds. This method is very painful to bulls. The bull may also undergo ataxia and fall down if the current is not controlled well.

- There is stimulation of sacral and pelvic nerves & leading to tetany condition to the bull. As the current is off; there is relaxed body of the bull which causes the ejaculation of semen in the form of dribbling.
- Semen is more in volume but may be contaminated because of prepucial hairs, dirt and bacteria etc i.e. quality is poor. Urine contamination may also be there.
- Pre-exposure of bull to estrous female, preputial washing and clipping of preputial hairs leads to better semen quality.
- This method is useful in crippled and weak bulls that cannot perform mounting and ejaculatory thrusts. Also, no dummy is required for the purpose.

4. ***Artificial vaginal (AV) method***

- Bull donates the semen in an artificial vagina because the artificial vagina mimics the natural vagina in terms of temperature, pressure and moisture. Temperature is more important factor than pressure for semen ejaculation in bull.
- It is the most commonly used method of semen collection in bulls. Semen quality is comparable with natural semen.
- Artificial vagina is made up of:
    - Outer hard rubber cylinder (35-40cm long with 8-10 cm diameter). It is smaller (25-35 cm for buffalo bull)
    - Inner latex/rubber liner (60-70 cm long with 7-7.5 cm diameter). It may be smooth (for younger bulls) or rough (preferred in old bulls)
    - A valve (5 cm away from one end of the outer rubber hard cylinder) for pouring warm water (temperature) and air (pressure)
    - Latex cone (20 cm long)
    - A graduated semen collecting glass tube of 15-20 ml volume
    - Insulating bag/jacket (protection of collected semen from thermal changes and UV rays)

Collection area

- It should be very close to the semen processing lab. If it is too far then more sperm will die due course of time and also there is more mechanical damages to the spermatozoa.

- The collection area should be of 20X30 ft^2 for two collection crates.
- The floor should be free from dust. It should be made of concrete with checkers or have a base of mixture of lime stone and sand. It should be non slippery.

Preparation of bull

- He should be washed and dried properly (water at the perineal region of dummy and preputial region of the donor may be passed into the collected semen which may cause metal toxicity or osmotic shock to the sperm).
- Prepucial washing with a mild antiseptic solution like potassium per magnate or normal saline is indicated for better quality of semen.
- The bull apron should be tied just behind the forelegs. It prevents the touching of erected penis to the body of dummy during mounting and reduces contamination.
- The bull should not be given exercise on the day of collection.

Preparation of dummy

- Female should not be taken as a dummy. Especially female in estrous (bull will be reluctant to donate semen to non estrous female in future), pregnant female (abortion may take place) or recently calved female (uterine infection and lochial discharge) should not be used as dummy. Bull should be trained to donate semen on male (preferably of the same breed) or non-estrous female.
- The dummy should be docile in nature and familiar to the donor bull. Any fighting or kicking may damage donor bull physically as well as psychologically.
- The dummy should neither be too heighted (donor bull poorly grasped the dummy and may fall after ejaculatory thrust) nor be too short (the dummy may sit or fall and may get physical trauma like fracture during mounting and thrust) in comparison to the donor bull.
- The dummy should also be washed, brushed and dried properly. The tail should be tied at one side before semen collection.

Preparation of semen collector

- Semen collector should be familiar with the bull. He should not wear flashy coloured clothes. He should wash his hands with soap or 1% savlon solution properly.
- He should not use abusive language or a punishing attitude to bull.

Preparation of AV (Artificial vagina)

- AV should be washed and sterilized in an autoclave. It should be capped with aluminium foil on open ends and placed in an incubator till semen collection.
- Pre-warming of AV with 70^{0}C water is done and then discard this water and fill warm water to maintain the temperature of AV **at 39^{0}C to 41^{0}C**. High temperature of AV may damage the penile mucosa and low temperature may lead to no ejaculatory thrust by the bull.
- Pressure is created in AV by blowing air through the valve (55 mmHg). The star like appearance of inner linings indicates the proper pressure.
- AV should be lubricated with non-irritating, non-toxic petroleum jelly like K-Y jelly. It can be applied with a sterilized glass rod up to a depth of 8-10 cm in AV.
- A rubber cone and semen collection tube should be installed at the side of near to valve. At the time of collection, the penis should be inserted from the end away from the valve so that in any instance, if penile holding is not proper, the penis would not be thrust and get damaged due to valve.

Semen collecton

- Semen collection should be done early in the morning. In the morning, emptied stomach bull is active and alert while he becomes lethargic after feeding and may be refused to mount. There is less environmental stress and noise because late in the day, sun & increasing environmental temperature is not suitable for a bull to donate the semen. Also, it provides more time to assess the seminal quality and accordingly early decision for 2nd collection or further seminal processing.
- The semen collector may whistle or produce soft voice signals (as bull has got training) at the time of bull approaching to dummy and mounting.
- The semen collector should stand on the right side of bull with AV in his right hand (to manage the strong jerk at the time of ejaculatory thrust)

so that he can easily hold the penile base (penis should not be touched) with left hand and direct it into the AV.

- Bull approaches to dummy and shows Flehman's reaction. 2-3 false mountings are recommended before collection. False mounting leads to more testosterone production with increased tubal activities of genitalia with improved libido. It will lead to improved seminal quality.
- As the bull mounts on a dummy and search for the vagina, semen collector should apply the AV (45^0 angle to the bull). He should take care of any trauma to the penis while penile insertion in to AV and ejaculatory thrust. Bull gives strong thrust and there is ejaculation into the AV.
- Pressure should be removed by loosening the valve and AV should be held verticaly in position after collection so that all semen drain and collects into the collection tube.
- Immediately semen should be kept in 37^0C water bath and transferred into the lab for evaluation and further processing.
- Fresh AV is used for each ejaculation (successful or unsuccessful semen ejaculation). AV, bull apron, Barn apron and gum boot of semen collector etc. should be washed in detergent solution immediately after collection and sterilized properly.
- 2-3 collections per bull per week are recommended. There should be 2 ejaculations per collection at the interval of 10-15 minutes. In a year 90 collections or 180 ejaculates from a bull is recommended.
- High frequency of semen collection leads to poor quality of semen (more immature sperms) and collection after long sexual rest also leads to more dead sperms in ejaculates.
- Sometimes there may be "**empty thrust**" i.e. thrust but no ejaculation (may be due to improper AV preparation or its application during collection) or "spilling" i.e. ejaculation without thrust.

Semen Collection in Ram/Buck

- A maximum of 20-30 seminal collections per day can be taken but to get better semen the quality it should be maximum 2-5 collections per day. Semen collection frequency is higher in ram than buck.
- Semen can be collected by:
    - Directly from vagina

- Electro-ejaculatory method
- AV method

1. ***Directly from vagina***

- Vulva should be cleaned before and after natural mating. Semen can be collected from the vagina with the help of a long pipette with a rubber bulb or nozzle fitted with a syringe.
- The Semen is of poor quality due to contamination with estrual discharge.

2. ***Electro-ejaculatory method***

- It is better to use this method of semen collection in the breeding season.
- The rectum is emptied and washed (enema) with normal saline or warm water. The animal is restrained in recumbent position (may be in standing position). Penis is exposed out of the preputial sheath and held with gauze. The urethral process or penile tip is inserted into the collection vial.
- A lubricated rectal probe is the inserted into the rectum up to 15-20 cm and applied at floor of pelvis with an AC current of 0V to 8V and then 8V to 0V in 5 seconds. This process is repeated for 3-4 times at the interval of 3-5 seconds. There is stimulation of the sacral nerve plexus and rhythmic stimulation of ampullae.
- Semen is ejaculated in the form of dribbling which is more in volume but with less sperm concentration. Semen may be contaminated with urine.

3. ***AV method: (Principle and method is same as in bull)***

- Female in estrous is preferred as a dummy as she will remain to stand quietly. Non estrous female or male can also be used.
- AV of length 20 cm with a diameter of 5 cm is used for semen collection in small ruminants.
- The temperature of AV should be **42^{0}C to 45^{0}C for ram** and **36^{0}C to 42^{0}C for buck**. Temperature is more important factor than pressure for semen ejaculation.
- In small ruminants, mounting and ejaculatory thrust is comparatively quick so semen collector should be alert.
- 2-3 ejaculations per collection is recommended.

Semen Collection in Boar

- In boar, semen is ejaculated in three fractions:
    - Prespermic fraction- Less/no/dead sperms with heavy contamination Gelatinous material from cowpers gland (20-30% of whole semen)
    - Sperm rich fraction- Creamy white in colour with rich sperm concentration (≈150 ml)
    - Postspermic fraction- Clear fluid of seminal plasma with little gel (≈200 ml) and no spermatozoa
- Less than 3 collections per week are recommended. One collection takes about 5-15 minutes.
- Semen collection of one boar in front of another boar stimulates the other boar (increased libido) for semen donation.
- Pressure is more important factor than temperature for semen ejaculation in boar.
- Semen can be collected by:
    - Gloved hand method
    - AV method

1. *Gloved hand method*

- It was discovered by Nova and Macpherson.
- Holding the erected cork screw shaped penis with a gloved hand manually with pressure mimics the natural vagina and cervical lock. The boar should not be touched with the gloved hand. Gloves should be sterilized and of non spermicidal material like vinyl nitrite (not latex).
- Dummy or phantom or estrous female is used. On stimulation, the penis protruded out which is held by a gloved hand applying little pressure. Pelvic thrusting is there and then at the time of ejaculation thrusting stops. Some more manual pressure is applied continuously till the completion of ejaculation (5-10 minutes).
- Multi-layered gauze or nylon filters are used over the collection pyrex bottle or sterile flask which separates the gel portion to be mixed with sperm rich fraction of the semen.

2. ***AV method***

- AV of 12.5 cm in length with a diameter of 4.5 cm is used. Temperature of AV should be **45°C to 50°C.**
- The sponge is used in between the outer cylinder and inner liner to maintain pressure on cork screw penis during ejaculation. Some-times rubber foam is used at the entrance of AV to avoid prepucial insertion into AV.

Semen Collection in Stallion

- Stallion ejaculates semen in three fractions:
  - Pre-spermic fraction -Watery seminal plasma without sperms
  - Sperm rich fraction - Milky white with high sperm concentration
  - Post-spermic fraction- White gel material (seminal vesicle & cowpers gland) without sperms
- Semen can be collected in stallion-
  - Condom method
  - AV method
  - Manual manipulation

1. ***Condom method***

- A sterilized condom is tied over erected penis of the stallion and let him perform natural mating.
- Semen is collected from condom after untying it from the penis post mating.

2. ***AV method***

- It is the most commonly used method for semen collection in stallion.
- Different types of AVs are used e.g. Cambridge, Russian, Missouri and Nishikava type etc
- An estrous female, a female injected with estrogen or phantom is used as a dummy. She should be controlled with hobbles. The Dummy should be cleaned and dried. The dummy tail head should be tied with a bandage before mounting. Hairs of the tail may cause irritation or inflammation in the vagina of dummy or penis of the stallion.

- Semen collector should use helmet and gum boots as protection.
- The stallion should be washed, brushed and dried properly. The penile part is washed with normal saline to remove smegma. It should be dried with a towel to avoid any water drops or soap remnants.
- AV should be sterilized with 70% isopropanol. AV for the stallion is large in size so provided with a handle. Its temperature should be maintained at **45°C to 50°C** for collection of semen.
- The Semen collection bottle is provided with a nylon filter to avoid mixing of gel material with semen.
- The stallion is teased with female in estrous or wooden dummy may be used. The after mounting, the penis is directed into AV. Ejaculation continues with thrusting and is completed in about 15-20 seconds.
- During ejaculation, there is peristaltic movement (**ejaculatory waves**) on the ventral surface of penis can be observed. Completion of ejaculation is indicated with flagging of the tail.
- After ejaculation, valve is released to remove the pressure and water from AV so that all the seminal content is drained and collected in the collection bottle.

Semen Collection in Dog

- Dog ejaculates semen in three fractions:
    - Pre-spermic fraction - Prostatic fraction; watery/cloudy seminal plasma without sperms (from urethral mucosa and gland of littre) (0.5 to 20 ml)
    - Sperm rich fraction - Milky white viscous fluid with high sperm concentration (0.5-4 ml)
    - Post-spermic fraction - Prostatic fraction; clear watery without sperm (3-30 ml)
- Total semen volume may be 1 to 40 ml per ejaculate.
- Pressure is more important factor than temperature for seminal ejaculation in dog.
- Semen collection frequency should be 2-4 times a week best quality. Two semen ejaculates can be collected at the interval of 30 minutes at every semen collection.

- Semen can be collected in dog by-
  - Manual manipulation method
  - AV method

1. ***Manual manipulation method***

- Teasing can be done with exposure to estrous female. Sometimes without female, dog can be stimulated with mild masturbation or massaging in the penile area. Vigorous or prolonged massaging may lead to bleeding of penile mucosa.
- Penis is exposed by retracting back the prepuce. Semen collection should be done after the complete erection of the penis and cessation of the trusting moment.
- Moderate pressure is applied on bulbus glandis (base of the penis) during pelvic thrusting or masturbating. Pressure is increased and continuously applied during seminal ejaculation (5-15 minutes).
- Semen is collected into a glass tube/vial with the help of a glass/plastic funnel.
- The second fraction of ejaculate should be tried to collect if the semen is to be subjected for a freezing process

2. ***AV method***

- AV of 15-20 cm in length and 5 cm diameter is used.
- The temperature of AV should be 40^{0}C to 42^{0}C for semen collection.
- Exposure to estrous female (teaser) can stimulate the dog.

Urination is not allowed before or during ejaculation. Precaution should be taken in long hair breeds so that hairs should not enter into AV during collection. Also AV should be removed gently after collection as forceful removal may cause laceration or superficial damage to bulbus glandis because of pressure.

Semen Collection In Tom

- By artificial vagina (44^0 to 46^{0}C) and electroejaculatory method (after anaesthesia) are used to collect the semen.

11

Evaluation of Semen Quality

Ajit Singh Mahla and Vikas Sachan

- Mature sperm cells are undifferentiated cells which are the testicular product as well as indicators of testicular spermatogenic function. A spermiogram reflects the testicular activity and fertility status of a male animal.
- Disease conditions, age, stress factors, vaccination or any altered physiological conditions may affect the seminal quality.

Objectives of Semen Evaluation

- To provide essential information about the fertility status of the male animal, though the fertilizing capacity of the animal cannot be completely defined by these tests.
- To diagnose various pathological conditions of testes and male genitalia.
- To determine the eligibility for further processing of semen cryopreservation
- To determine the potential of semen for AI purposes

- The seminal examination should be done with the following considerations
  - At least 2 ejaculates should be analysed with basic tests at an interval of 7 days.
  - There should be no urine contamination (to avoid pH change)
  - There should be no mixing with water or any other solvent (to avoid osmotic shock)
  - There should be no vigorous handling (to avoid mechanical shock)
  - There should be no contamination with dirt, dust, dung etc.
  - There should be no exposure of neat semen to sunlight (to avoid photo-oxidative damage)

- Evaluation should be done with minimum delay (to avoid deterioration of the semen quality)
- The semen sample should be kept in a water bath at 32^0 to 36^0 C temperature (to avoid cold or heat shock)

- Seminal examination can be categorized as:
  1. Gross examination (macroscopic or physical examination)
  2. Microscopic examination
  3. Biochemical examinations
  4. Bacteriological tests
  5. Fertility examination
  6. Other supporting tests

1. Gross Examination

- Seminal volume, colour, consistency, density, odour, pH or any abnormality are examined grossly. These examinations of the seminal quality at the initial stage help in determining the breeding soundness of the bull. These examinations should be completed within 8-10 minutes.
- Semen should have a uniform, opaque appearance, free from preputial hairs, dirt, dung, pus, urine and lubricant material. Dung and dirt can be seen as they usually accumulate at the bottom of the tube. Foam at the top of semen is indicative of normal ejaculation which may disappear after some time.

1.1. Seminal volume

- The seminal vesicle and prostate gland contribute maximum in seminal volume along with a little contribution from the epididymis and Cowper's gland.
- Reduced seminal volume may be found in case of poor libido, testicular inflammation, obstruction in the ejaculatory duct, very young or old bull, poor teasing of a bull before collection, $3^{rd}/4^{th}$ ejaculate from the same bull and excessively used (frequent semen collection) bull etc.
- Increased seminal volume may be obtained due to various factors like second ejaculation, more body size, uro/pyo/haemospermia and more teasing etc.

- Indigenous zebu bulls have lesser seminal volume (3-6 ml) than exotic breed bulls (8-10 ml). Average semen volume (ml) per ejaculate in different species –

S.No.	Species/Animal	Seminal volume (ml)
1	Cattle Bull	5 to 8
2	Buffalo bull	3-5
3	Ram / buck	0.6to 1 ml
4	Boar	150-200 ml
5	Stallion	60-100 ml
6	Dog	1-25 ml
7	Cock	0.3 to 0.7 ml

- Seminal volume is not accurately measured with a semen graduated tube because there are huge numbers of spermatozoa engaged in bubbles at the top of the seminal surface. It is better to allow the froth to settle down before the estimation of volume. Nishikawa was the first to establish that weighing of ejaculate may give a more accurate estimation of semen volume (the specific gravity of bull ejaculate is about 1.036).

1.2. Semen colour and odour

- Semen should be collected in a glass tube and observed in natural daylight.
- Cattle Bull semen is usually yellowish white to creamy while it is whitish in buffalo bull. In ram and buck, it is creamy in colour.

S.No.	Colour of semen	Condition
1	Yellowish to Creamy white	Normal
2	Pinkish/Reddish/Brownish	Inflammation of accessory sex glands, Bleeding in genitalia (haemospermia)
3	Brown/blackish	Fecal contamination
4	Yellow	Pus (Pyospermia)
5	Watery yellow	Urine (Urospermia)
6	Yellowish green	Pseudomonas aeruginosa infection, purulent degeneration (genital pathology)

- The normal yellowish colour of semen is due to riboflavin and flavins from seminal vesicle & melanin or lipochrome from ampullae.
- The peculiar smell of the semen is because of prostatic secretions. Any abnormal/foul/fishy odour in semen may be due to purulent degeneration or any genital pathology (bacterial infection).

1.3. Consistency and density

- Seminal ejaculate should be uniformly opaque in appearance. Non-uniform opaqueness or flakes indicate the infection of genital organs. Density along with consistency is a rough qualitative predictor of sperm concentration.
- Consistency may be thick creamy to watery according to the sperm concentration. Pathological conditions of the testis, epididymis or accessory sex glands affect the consistency of the ejaculates.
- Sperm concentration in bull ejaculate should be ≥500 million per ml of semen for insemination.

S.No.	Consistency	Sperm concentration	Density	Grade
1	Thick creamy	≥ 2000 million/ml	DDDDD	Excellent
2	Creamy	1500-2000 million/ml	DDDD	Very good
3	Thin creamy	1000-1500 million/ml	DDD	Good
4	Milky	500-1000 million/ml	DD	Poor
5	Watery to mild cloudy	≤ 500 million/ml	D	Very poor

1.4 pH value

- Indicator of the presence of hydrogen ion concentration in the ejaculate i.e. balances between alkaline secretion of seminal vesicle and acidic secretion of prostate gland along with secretions of other accessory sex glands. Spermatozoa convert fructose into lactic acid during metabolism and release H^+, hence pH of a good semen sample of a bull is slightly acidic.
- It can be measured with a B.D.H. capillator (bromothymol blue or bromocresol purple reagents), pH paper strip (narrow range of 6 and 8.5) or pH meter (best).
- Normal pH range of semen in different species -

S.No.	Species/Animal	pH range
1	Cattle bull	6.5 to 6.9
2	Buffalo bull	6.7 to 7.0
3	Buck and Ram	6.4 to 7.0
4	Dog	6.4 to 6.8
5	Horse	7.2 to 7.7
6	Boar	7.3 to 7.9
7	Camel	7.4 to 8.6

- pH elevation (alkaline side) indicates bacterial infection (release of ammonia), more dead sperms (necrozoospermia), low sperm concentration (azoospermia or oligozoospermia), disease condition and excessive use of bull etc. A delay in the seminal evaluation may lead to lower pH.

1.5. Leucocyte test

- The presence of leukocytes in semen can be tested by a reagent like California mastitis reagent (gel formation in 15 sec). leukocyte percentage can also be accessed by staining the seminal smear with blue methylene stain.

1.6. Osmotic pressure

- It is necessary to maintain the osmolarity to prevent exosmosis or endosmosis through a semipermeable membrane of the sperm cell in the seminal fluid which maintains the integrity of the cell.
- The osmotic pressure of semen samples (bull semen) is about 280–300 mOsmol. It can be measured by an osmometer.

2. Microscopic Examination

- Mass motility, progressive motility, sperm concentration, % live spermatozoa, sperm morphological abnormality, % intact acrosome sperms, presence of other cells and sperm membrane integrity etc. are good predictors for seminal quality.

2.1. Sperm motility

- It is the most common parameter to predict sperm quality, especially at the field level. It also predicts the future fate of the sperm quality for AI or cryopreservation.
- All motile sperms in a semen sample are live but all live sperms are not motile.
- The epididymal spermatozoa are motionless, but after coming in contact with the accessory gland secretions, they gain motility.
- Frequency of semen collection, season, age, faulty collection, contamination, and pathological conditions affect sperm motility.
- Well-mixed semen and microscope with biotherm stage 36^0-38^0C are essential for motility assessment.

Mass motility – (Gross/initial motility)

- It is the motility of the whole sperm population in form of wave (collective movement of spermatozoa) which is produced because of the live progressively motile spermatozoa.
- It provides a rough prediction of sperm concentration and viability as some weak motile/non progressive motile or dead sperms are also dragged under the influence of progressive motile sperms.
- Mass motility is observed immediately after collection in the fresh semen sample. A drop (not too small or too big) of fresh undiluted semen is placed on a pre warmed (37^{0}C) slide without cover slip and observed under 10X (low power microscope). Observations should be avoided near air drop, if any.
- Grading of mass motility of a semen sample:

S.No.	Gradings	Observations and findings
1	5; Excellent	Extremely rapid eddies and waves (origin & disappearance cannot be traceable); Individual sperms couldn't be seen;80-100% spermatozoa are progressively motile
2	4; Good	Less rapid eddies and waves; Swirls move toward extremities; Individual sperms could not be seen; 60-80% spermatozoa are progressive motile
3	3; Fair	Slow eddies and waves; Individual sperms could be seen hardly; 40-60% spermatozoa are progressive motile
4	2; Poor	No eddies and waves; Individual sperms could be seen (Oscillating & circulating); 20-40% spermatozoa are progressive motile
5	1; Very poor	No eddies and waves; Individual throbbing sperms could be seen; 10-20% spermatozoa are progressive motile
6	0; All dead	No motility.

- It is a qualitative estimation of motility so sometimes grading can also be done with the fraction of 0.5 e.g. 0.5, 1.0; 1.5, 2.0, 2.5, 3.0, 3.5, 4.0, 4.5, 5.0. The semen sample with mass motility of more than 3 is considered fit for preservation or insemination purpose.
- Size, depth of drop and temperature influence the observations. Poor mass motility is seen in case of long sexual rest, more dead sperms, thermal shock (cold/hot), hot &humid climate, chemical exposure, dirty glassware, and contaminations etc.
- Low sperm concentration e.g., in the case of semen collected through electro-ejaculation method may not show rapid waves even with very good progressive sperm motility because of poor sperm concentration.

- "**Static ejaculates**" are one in which mass motility is zero or approximately zero but after dilution with an extender at 37°C sperms revive and gain motility. This is the character of buffalo bull semen. Such semen sample has poor freezability and fertilizing ability.

Individual motility – (Progressive motility; %)

- It is the observation of the motility of individual sperm cell. % progressive motility is the estimation of spermatozoa that are progressive motile (movement in a progressive direction). Individual motility can be predicted only in diluted semen sample. Normally speed of a bull sperm cell is 60 to 162 (120) µm/sec.
- % Progressive motility is the best predictor of sperm quality at field level which can indicate the fertilizing capacity of spermatozoa.
- A drop (10 µl) of diluted semen (1:100 or more) is taken on clean grease free pre warmed (37°C) glass slide and is covered with a cover slip (depth of 20µm for free swimming spermatozoa). Observations are taken under 20X to 40X microscope.
- Change in pH, osmolarity and temperature may affect the progressive motility.
- Various types of individual sperm movements can be observed -

S.No.	Type of movement	Reason
1	No movement	Dead sperm
2	Forward	Progressive motile sperms
3	Circular	Cold shock or osmotic shock
4	Reverse circular	Cold shock/osmotic shock or bent tail
5	Throbbing/oscillatory	Aged sperm or energy deficit

- Grading of semen sample on the basis of % progressive motility of sperms -
  - Excellent - ≥ 80%
  - Good - 60-80%
  - Fair - 40-60%
  - Poor - 20-40%
  - Very Poor - ≤ 20%
- A semen sample with ≥50% progressive motility can be used for insemination and with ≥ 60% progressive motility can be used for the freezing process.7

- The Heat method (50^0 C for 5 min), 1% chlorazine or haemocytometer method can also used to predict the progressive motility. In these methods non-progressive motile sperms (in non treated sample) are counted and then total sperms are counted after killing them (treated). The difference in these counts results in progressive motile spermatozoa.
- Motility in dog semen can be assessed up to 21^{0}C (37^{0}C is not necessary) as canine sperm is comparatively resistant to cold shock.
- CASA (computer assisted semen analyzer) is also used for determining the individual motility with more precision.

2.2. Sperm viability (% live spermatozoa)

- The sperm membrane is involved in metabolic changes, capacitation, acrosome reaction and the binding of the spermatozoan to the oocyte i.e., important for the fertilizing capacity of spermatozoa.
- Sperm viability is determined by motility or by plasma membrane integrity (for non motile but live sperms). As membrane integrity is the key component to describe the viability of spermatozoa, max assays that are used for viability determination are based on the functionality (permeability) of plasma membrane.
- For evaluation of cell viability based on cell membrane integrity, HOS test is considered better than eosine-nigrosin stain as eosine-nigrosin reflects the structural integrity of the cell while HOS test reflects the functional integrity of the cell membrane.
- Sperm viability should be accessed within 30 min to 1 hr of fresh ejaculation.
- Per cent viable sperms are always more than Per cent progressive motile sperm because all progressive motile sperms are live but all live sperms are not progressively motile. Per cent viable sperms are more in bull above 3 years of age, 1st and 2nd ejaculates and in the season other than hot and humid.

a. Supravital staining – (Eosine-nigrosin satin and trypan blue stain)

Eosine-Nigrosin stain

- Principle – Eosine cannot penetrate the cell membrane of live sperm due to maintained selective membrane permeability which result in a colourless appearance (Glistening white) but stain penetrates the membrane of dead sperms (lost character of selective permeability) resulting in a pink appearance. Nigrosin imparts a dark background for better visibility.

- Eosine-Nigrosin stain should be isotonic to semen. Osmolarity can be adjusted by adding sodium citrate. pH of the stain should be ranges from 6.8 to 7.0. If contamination is avoided, the prepared stain can be stored for up to 6 months at room temperature and even for years in refrigerated condition.
- A drop of semen is taken on a clean greese free slide and 2-3 drops of stain are added. After gentle mixing, it is left for 1 to 1.5 minutes. A smear is made and air-dried following mixing the semen and stain gently and leaving for 1 to 1.5 minutes. A smear is prepared and air dried. Observation is taken under 100X power with immersion oil. At least 200 sperms are counted (partial or complete pink colour-stained sperms are considered dead and glistening white or unstained sperms are counted as live).
- Eosine-Nigrosin staining is better for fresh semen as compared to thawed semen to assess spermatozoa viability as glycerol interferes with the stains. Glycerol increases the cell permeability and some live cells appear pink stained following eosine incorporation.
- Bulls having $\geq$ 70% live sperms in ejaculate are considered to be good fertile. These samples may also be subjected to freezing process. Thawed semen samples having >50 % live spermatozoa are considered good for insemination.

b. *Fluorescence techniques*

- Some fluorescent stains can also be used for estimating sperm viability e.g., Hoechst (H) 33342, H 33258, DAPI/PI (4',6-diamidino-2-phenylindole/propidium iodide), SYBR-14/PI etc. These works are based on their permeability to sperm cells and the production of fluorescence.
- Hoechst (H) 33342, H 33258 are DNA specific stains. Dead sperms DNA are stained easily with Hoechst stain and produce blue/cyan fluorescence. SYBR-14 and DAPI are also DNA specific stains which can cross the plasma membrane of both live and dead sperm but PI is cell impermeant DNA stain i.e., PI can cross only dead sperm. Live spermatozoa stained with SYBR-14 give fluorescence of green color while with DAPI, they give blue color fluorescence. Dead sperms give fluorescence of red orange with PI stain (DNA has more affinity for PI in comparison to DAPI or SYBR14).

S.No.	Stain	Sperm cell type	Fluorescence color
1	Hoechst (H) 33342, H 33258	Live and dead	Blue/cyan
2	SYBR-14	Live and dead	Green
3	DAPI	Live and dead	Blue
4	PI	Dead sperm	Red orange

c. ***Membrane integrity – (Hypo-osmotic swelling test, HOS test) (%)***

- Jeyendran and Cowrkers (1984) developed the HOS test to evaluate the functional integrity of the sperm cell membrane of human spermatozoa.
- **Principle** – When a sperm cell is subjected to a hypo-osmotic solution, because of the endosmosis i.e. water influxes into the intracellular space through a biologically active plasma membrane (live sperm plasma membrane maintains its semipermeable or selective permeable character). The Sperm plasma membrane swells up especially at tail region, and curling appears to various degree. The effect is more in the tail because the cell membrane in this region is thinner and loosely attached.These changes extend from the distal tail end to mid piece and head on longer exposure.
- The osmotic pressure of the bull seminal plasma is 280 to 300 mOsm. Generally, a 150 mOsm, hypo-osmotic solution (Sodium citrate, fructose, and distilled water) is used. It causes the influx of water into the cell without cell lysis.
- 0.1 ml of freshly collected undiluted semen is mixed with 0.9 ml of hypo-osmotic solution (150 mOsm) in a small test tube at 37° C. The suspension is incubated at 37° C for 60 minutes. A very meagre amount of nigrosin can be added for better visualization. A drop is placed on glass slide and observed under 40 X microscope. At least: total 200 sperms should be counted in different fields.
- Sperms with swelling and curling of various degrees in the tail are considered as HOS positive sperms i.e., live sperm with membrane biochemical integrity. Spermatozoa with curled tail in control solution (normal saline) are subtracted from total spermatozoa with swelling and curled tails in hypo-osmotic solution. The Percentage of the difference is interpreted as HOS + sperms in the given semen sample. A drop of eosine solution can also be added to detect the dead sperms and all pink stained spermatozoa can be excluded from the count of HOS-responsive spermatozoa.
- More than 60% HOST positive sperm in fresh semen is considered as good for insemination or cryopreservation. Post thaw semen with ≥ 40 % HOST positive sperms can be used for insemination.

- Distilled water may also be used in place of hype-osmotic solution and is termed as "**water test**".

d. ***Lipid peroxidation***

- The level of lipid peroxidation (lipid extraction and lipid peroxidation) in sperm plasma membrane can also be quantified. It indicates the membrane integrity and viability of the sperms.

2.3. Sperm concentration

- It is one of the most important parameters to predict the quality of semen in terms of the fertilizing ability of the bull. It is also essential for determining the dilution rate in the process of semen dose production in semen laboratories which greatly affect the production potential.
- A Rough qualitative estimation of sperm concentration in a semen sample can be done by observing its consistency. Other methods for quantitative estimations are: Neubauer's haemocytometer, Spectrophotometer, Photoelectric colorimeter method, Opacity tube method, Electronic sperm cell counter, Spermometer (Makler sperm counting chamber), CASA and flow cytometer etc.

a. ***Neubauer's haemocytometer method***

- Principle – On dilution with semen diluting fluid (for proper dispersal of the cells), spermatozoa become dead due to formalin and become non motile. These non motile sperms are counted easily after loading in Neubauer's chamber.
- Semen diluting fluid is prepared by sodium citrate, eosine yellow, formalin, and distilled water. Eosine improves the visibility of the sperm cells.
- Generally, the dilution rate for bovine, ovine/caprine and equine are 1:100, 1:200 and 1:20 respectively. Dilution rate can be altered depending on the sperm concentration in the ejaculate. For dilution purpose and loading the neubauer's chamber, RBC or WBC pipette are generally used.
- Sperms are counted in 5 tertiary squares (RBC chambers). If the counting of sperms is "N", then the sperm concentration in the semen sample is 10N million spermatozoa/ml, provided that the dilution rate is 1:200.
- Sperm concentration (million/ml)

S.No.	Species/Animal	Sperm concentration (million/ml)
1	Cattle bull	1200-1800
2	Buffalo bull	800-1200
3	Buck/Ram	2000-6000
4	Boar	200-300
5	Stallion	120-300
6	Dog	100-700
7	Cock	3000-7000

b. A spectrophotometer or calorimetric method is used for calculating sperm concentration in a given semen sample based on optical density. The optical density (more for more sperms per unit volume) of the sample is measured against the calibrated optical density of the known sample. Contaminated sample (debris, egg yolk etc)and diluted semen reduces the accuracy by spectrophotometer.

c. In sperm counting chamber like Makler chamber, a semen sample is loaded and covered with coverslip with grids (10 X 10). Sperms are counted directly in any straight 10 boxes and the obtained count is the concentration of sperm in million/ml.

d. CASA and ***flow cytometer*** are also used for determining semen concentration with more accuracy.

- Teasing before ejaculation, frequency of collection, premature ejaculation, and disease condition etc. may affect sperm concentration.
- Semen sample of a bull having spermatozoa more than 600 million/ml is used for insemination and cryopreservation process.

2.4. Sperm abnormalities or defects

- Different types of abnormalities may be responsible for poor sperm maturation process, reduced motility, abnormal sperm biofunctions or embryonic death even after ova penetration.
- Spermatozoal abnormalities are valuable assessments for determining the potential fertility of the bulls. Sperm cell with abnormal morphology is also as important as motility assessment because some normal motile sperms may have some defects like damaged acrosome or genetic defects which make the fertility potential of sperm questionable.
- Sperm morphology can be considered as one of the most reliable criteria for semen evaluation as it is influenced least by the collection process. It is better to examine fresh semen or semen diluted with physiological saline for morphological examination as egg yolk may hamper the observations.

- Giemsa staining technique is considered as one of the best staining techniques for sperm morphology. Eosin-nigrosin stain, Rose Bengal, Williams stain, Wright's-Giemsa stain, Papanicolaou, India ink, Casarett's stain, Leishman staining, Hematoxylin-Eosin-Giemsa stian, Carbolfuchsin-Eosin-Methylene blue etc are also used conveniently to assess the sperm morphological abnormalities.
- In fresh or thawed semen, total abnormalities should not be more than 20% (<5% abnormality of an individual class). Permissible major defect of any one type should be < 10% and minor defect of any one type should be < 20%.
- Sperm cell morphology can also be visualized quickly with a **wet mount test** without staining.

Cells or material other than spermatozoa in semen

- Leucocytes (infection and inflammation of genitalia) and erythrocytes
- Squamous epithelial cells (from preputial or urinary tract epithelium)
- Boat shape cells (spermatozoa with deformed head)
- Immature germ cells (spermatids, spermatocytes etc)
- Medusa cells (brush like ciliated epithelial cells from the efferent ducts; visible in testicular hypoplasia). These are more common in equines.
- Giant cells with multiple nuclei (undivided spermatocytes; observed in testicular hypoplasia or degeneration)
- Clusters of degenerated sperm cells
- Free floating cytoplasmic droplets

Spermiogram Terminology

S.No.	Parameter	Evaluation criteria	Nomenclature
1.	Sperm concentration	Normal	Normozoospermia
		Reduced	Oligozoospermia
		Increased	Polyzoospermia
		Zero	Azoospermia
2.	Seminal Volume	No ejaculation	Aspermia
		Reduced	Hypospermia
		Increased	Hyperspermia
3.	Sperm motility	Decreased (<50% sperm with progressive motility)	Asthenozoospermia

S.No.	Parameter	Evaluation criteria	Nomenclature
4.	Sperm viability	All dead	Necrozoospermia
5.	Abnormal sperm	>70% abnormal sperms	Teratozoospermia
6.	Contamination	With urine	Urospermia
		Pus cells	Pyospermia
		RBCs	Hemospermia

3. Biochemical Examinations

(Metabolic activity, Resistance to adverse conditions, Enzymatic activity)

Metabolic activity the spermatozoa - Spermatozoa produce energy mostly by anaerobic glycolysis and by oxidation of endogenous substrates. Spermatozoa utilizes lactate or pyruvate (resulting from anaerobic fructolysis) to produce ATP i.e. source of energy. Spermatozoa also produce energy under aerobic conditions (oxidation pathway in the mitochondria). The motility of spermatozoon is a direct indication of its metabolic activity.

- There are several factors like temperature, pH, concentration of sperms, cations/anions, osmotic pressure, hormones etc. that affect sperm metabolic activity.
- Semen sample with more sperm concentration shows lower metabolic activity per spermatozoon. Faster reduction of pH of a semen sample indicates better metabolic activity (more production of lactic acid).

3.1. Fructolysis index (FI) - (Mann, 1948)

- Fructolysis is a prime activity for the production of energy anaerobically (metabolic activity) by bull and ram spermatozoa.
- It is a test to estimate the metabolic activity (activity of intracellular enzyme dehydrogenase) of the sperm. Better sperm metabolism leads to a higher rate of utilization of fructose in a semen sample. It is the amount of fructose (mg) utilized (or lactic acid produced) by 10^9 bull spermatozoa in 1 hour at 37°C.
- Seminal fructose is incubated with acidic resorcinol, resulting in the formation of a chocolate brown or red colored compound. Its intensity which is observed with colorimeter (Optical density against standard) indicates the amount of utilized fructose.
- The Fructolysis index should be 1-2 mg for bull semen.
- The presence of other substrate like citrate, lactate, glucose etc. affect the fructose reduction rate.

- The Fructose disappearance rate (FD) is also used to estimate the metabolic activity of a semen sample. It is the reduction in fructose concentration in 0.1 ml semen in 1 hr at 37^{0}C.

3.2. Methylene blue dye (MB) reduction test – (MBRT)

- This test indicates the metabolic activity (dehydrogenase activity) of the sperms in a semen sample.
- More sperm metabolic activity leads to more liberation of H^+ under anaerobic conditions which reduces the methylene blue (blue colour) into leuco methylene blue (colourless). Lesser time taken in reduction indicates better metabolic activity (release more hydrogen ions per unit time) of sperm in a semen sample. It should be less than 5 min for cattle bull and less than 8 min for buffalo bull semen. Samples taking more than 9 minutes in reducing methylene blue are discarded.
- More time in MBRT indicates poor mitochondrial activity and poor fertilizing capacity.

3.3 Resazurine dye reduction test

- It is also a test to indicate the metabolic activity of the sperms. More metabolically active sperms produce more H^+ which reduces resazurin (blue) into resorufin (Purple/Pink in 1 minute) and further into dihydroresorufin (colourless in 3-4 minutes).
- It should be <5 min for a good quality semen sample.

S.No.	Methylene blue dye reduction time (min)	Resazurin dye reduction time (min)	Semen quality
1	3-5	3-5	Good
2	5-9	5-8	Fair
3	>9	>8	poor

- Quantitative estimations of methylene blue reduction and resazurin reduction rates can be done with spectrophotometric methods.
- Alteration in temperature, contamination, microorganisms and oxygen exposure may alter the MB reduction and resazurin dye reduction rate.

3.4 Millovanov's resistance test – (R test)

- It is the measurement of resistance of the sperm to the salt i.e. ability to withstand 1% NaCl.

- Progressive motility is observed after adding 10 ml of 1% NaCl at regular intervals of 10 minutes till the complete cessation of forward motility. R value is the amount of 1% NaCl solution (Referred to as R value) required to cease the progressive motility of sperm in 0.02 ml semen sample.
- The R value of a good quality semen should be > 5000.

  R value = Volume of 1% NaCl (ml) used / 0.02 ml (volume of semen sample)

3.5 Cold shock resistance test

- There are irreversible damages to the sperm (loss of ATP, loss of enzymes and various proteins etc.) during the freezing process at 5^{0}C which leads to a reduction in the survival and fertilizing capacity of spermatozoa.
- This test investigate the ability of spermatozoa to withstand the cold shock. Semen is placed in a beaker having crushed ice in the refrigerator at 5^{0}C and observation regarding motility (Progressive, Non-progressive and immotile) after every 10 minutes. A semen sample having motile sperm for a longer duration is considered of good quality.
- It indicates the probability of spermatozoa to undergo the process of preservation i.e., longer fertilizing ability and storage ability with better fertility.
- Second semen ejaculate has better resistance to withstand the cold shock than 1st one.

3.6. Post-thaw Incubation test

- The semen sample is thawed and incubated again at 37^{0}C. A sperm motility assessment is done every 15 minutes. A smaller drop in motility after each observation indicates a greater chance for the thawed spermatozoa to survive in the female reproductive tract ensuring better fertility.

3.7. Oxygen utilization test – (Walton & Edward, 1938)

- Sperm utilizes oxygen for exogenous and endogenous oxidation during respiration. More oxygen utilization per unit time by a semen sample indicates more sperm activity indicating better sperm quality. Oxygen utilization by a semen sample is measured with the help of a **Warburg respirometer**.

- On average; 100 million bull sperm utilizes about 21 **μL** O_2 per hour.
- Oxygen utilization by the sperms can also be expressed as respiratory quotient (RQ). The **Respiratory quotient** (RQ) of spermatozoa is the ratio of the volume of carbon dioxide produced by the volume of oxygen consumed in the given unit of time. It is expressed as $ZO_{2.}$ The average ZO2 value of bull semen is 21.

3.8. Ascorbic acid concentrations

- Ascorbic acid acts as an anti-oxidant for sperm cells. More ascorbic acid in semen indicates better protection of spermatozoa against oxidative damage (due to free radicles)and better quality of the semen.

3.9. Enzymatic estimations

***Glutamic oxaloacetic transaminase or aspirate amino transferase* (GOT/ AST)**

***Glutamic-pyruvic transaminase or alanine aminotransferase (GPT/ALT)* activity**

- These are intracellular enzymes located in midpiece of the sperm cell which release only if there is damage to the sperm cell. The presence of ALT and AST in semen sample indicates damaged or dead sperm cells and poor quality of the semen.

Hyaluronidase and acrosin activity

- These enzymes are integral components of acrosomal content. Its presence in extracellular fluid indicates the damaged acrosome. More is the level of **hyaluronidase** and **acrosin** activity in seminal fluid, the poorer the quality of the semen sample in terms of fertility.

Catalase test

- Catalase is an enzymatic antioxidant component found in the sperm cell cytoplasm.
- More catalase activity in the sperm cells is indicative of better quality of semen sample. It also indicates blood, pus, bacterial contamination, or dead cells in semen sample.

Alkaline phosphatase activity

- Semen phosphatases take part in dephosphorylation during sperm metabolism. It is produced in epididymis and reflects the functionality as well as metabolic activity of spermatozoa. Higher alkaline phosphatase in semen indicates better fertilizing capacity.
- In an azoospermic semen sample, a low ALP indicates ductal blockage, whereas a normal ALP indicates testicular failure. An incomplete collection can also result in low ALP.
- Other enzymatic estimations like phosphatase, lactate dehydrogenase, esterases etc. also indicate the seminal quality. Enzymes in seminal plasma like deoxyribonuclease-1 and tissue inhibitor of metalloproteinase (TIMP-2) are some more good indicators of seminal quality.

3.10. Buffering capacity

- Buffering capacity is the ability of semen to maintain the pH (least change) to absorb acid / alkali. Better the buffering capacity; better will be the survivability of spermatozoa in female genitalia.
- Bull semen :
  - Highly buffered at pH levels below 5.5 and above 9.0
  - Moderately well buffered at pH 5.5-6.5 and pH 8.0 to 9.0
  - No/poor buffering capacity between pH 5.5 to 8.0
  - The semen of dog and stallion is virtually without buffering capacity.

4. Microbial Examination (Bacteria, virus, protozoa, molds etc)

- Infection in the male genitalia, presence of pus cells and bacteria in the semen deteriorate the spermatozoa viability leading to poor sperm quality and low fertility i.e. poor sperm quality.
- Infectious microbes (venereal, systemic, or opportunistic) such as *Corynebacterium renale*, *E. coli*, *Corynebacterium pyogenes*, *Brucella abortus*, *Mycobacterium tuberculosis*, *Mycobacterium paratuberculosis*, *Pseudomonas aeruginosa*, Streptococci, Leptospira, Staphylococci, *Mycoplasma bovigenitalium*, Ureaplasma, genital fibropapilloma virus, FMD virus, IBR-IPV, BVD etc may be found in the semen of bulls.
- Semen may contain *Brucella ovis* in sheep, *trypanosome equiperdum* and *Infectious equine anaemia* virus in equine.

- The sample should be examined for bacterial diseases (Brucellosis, Campylobacteriosis, Leptospirosis, T.B.& J.D), Viral diseases (FMD, IBR, BVD, R.P) and fungal diseases. The semen should be free from all such diseases causing pathogens.
- Bacterial count assessment in semen sample can be done with the standard plate count method and expressed as colony forming unit (CFU)/ml. The Permissible bacterial load for frozen thawed semen is **<500 CFU/ml.**

5. Fertility Test

- Apparently good quality semen without any visible defects in sperms may also have poor fertility. The ultimate quality of semen should be determined with the help of fertility tests. Any molecular defect in spermatozoa leads to failure of its attachment to zona pellucida or failure of fertilization.
- In vivo fertility tests are time consuming as well as affected by multiple factors. In vitro tests are cost effective and less time consuming. Various *in vitro* fertility tests like zona binding assay (ZBA), hyaluronic binding assay (HBA), hemi-zona binding assay, *in vitro* fertilization (IVF) etc. are meant to determine the fertilizing capacity of the spermatozoa.

5.1. Zona binding assay (ZBA)

- The zona of a mature ova has receptors like ZP1, ZP2, ZP3 etc. which allows the spermatozoa to bind on zona. The binding of spermatozoa on zona pellucida is a good indicator of sperm quality in terms of morphology, motility, and acrosomal integrity etc. More number of sperms attached to zona pellucida of exposed ova indicates better quality of semen.
- Most commonly, due to species specificity, homologous oocytes are used. Heterologous oocytes of some species may show zona binding with spermatozoa of some limited other species like equine sperms can attach to zona of porcine and bovine oocyte. Porcine spermatozoa may also bind with oocyte of equines.
- As general protocol:
  - Ova are collected (in PBS or NS) and they are made cumulus free
  - Washed with fert TALP or mBWW(modified *Biggers-Whitten-Whittingham*)

- Thawed sperms are diluted (0.1 million motile spermatozoa/ml)
- Co-incubation of 50µl of sperm suspension with 8-10 oocytes in CO_2 incubator (5% CO_2, 95% humidity, 37ºC for 4 hrs)
- Oocyte-sperm complexes are then subjected to washing with PBS or fert-TALP, fixing with Glutaraldehyde or paraformaldehyde and staining with DNA specific stain e.g. Hoechst 33342)
- Observation under fluorescent microscope (Blue emitting fluorescent)

- At least 200 sperms should be attached with one oocyte to be considered of good quality. The quality of the test semen sample can be also compared with the control one in the form of zona binding index (ZBI).

$$ZBI = \frac{No.\ of\ spermatozoa\ bound\ to\ oocyte\ (test\ semen\ sample)}{No.\ of\ spermatozoa\ bound\ to\ oocyte\ (control\ semen\ sample)}$$

5.2. Hemi Zona assay (HZA)

- The principle and procedure are almost the same as the zona binding assay.
- To manage the problem of shortage of oocytes, one oocyte is bisected into two halves. Bi-sectioning of the oocyte is performed with the help of a micromanipulator under a zoom stereomicroscope or phase contrast microscope.
- The ooplasm of each half ova is removed and then co-incubated with spermatozoa in a CO_2 incubator. Sperms attached to the outer surface of the zona are considered for interpretation. Test and control semen sample quality can be assessed comparatively and expressed as the "Hemizona index".

$$Hemi\ zona\ index = \frac{No.\ of\ sperms\ bound\ to\ outer\ surface\ of\ hemi\ zona\ (test\ sample)}{No.\ of\ sperms\ bound\ to\ outer\ surface\ of\ hemi\ zona\ (test\ sample)}$$

5.3. Zona-free hamster ova penetration test – (sperm penetration assay)

- The best essay to evaluate the fertilizing capacity of a male animal would be to study the spermatozoa penetration to homologous species eggs or zona-free ova of heterologous species egg like hamster.
- Zona-free ova of most mammalian species, except hamster, are species specific and reject the entry of spermatozoa of most other species. Zona-free hamster eggs allow attachment and penetration of capacitated

sperms from almost all species of animals. Sperm interaction with vitelline membrane reflects the conception rate.

- General procedure/steps involved:
  - Superovulation (PMSG and hCG) and collection of ova (from hamster ovary)
  - Washing with sperm TALP or mBWW
  - Make ova cumulus free (0.1 % hyaluronidase)
  - Make ova zona free (0.1% trypsin)
  - Preparation of sperm suspension (3-6 million/ml)in mBWW media and capacitation media (Ca ionophores)
  - Co-incubation of 8-15zona-free hamster ova with 100 µl capacitated sperm suspension in a CO_2 incubator (5% CO_2) at 39°C for 3 hrs.
  - Aceto-orcein staining of sperms-oocytes complex (Aceto-orcein is DNA specific which gives a florescent colour of purple-red)
  - Examination of ova under 400 X magnification
- Sperms with decondensed heads with tails are considered to have penetrated the ova. The semen sample is considered of good quality if 10% of oocytes are penetrated with sperms (@ 5 sperms per oocyte).
- The fertilization capacity of sperms is determined in terms of fertilization index (FI) or fertilization % (FP).

$$Fertility\ index = \frac{No.\ of\ swollen\ sperm\ head}{No.\ of\ oocyte\ inseminated}$$

$$Fertilization\ percentage = \frac{No.\ of\ oocyte\ penetrated}{No.\ of\ oocyte\ inseminated} X\ 100$$

5.4. Bovine cervical mucus penetration test (BCMPT) or sperm mucus penetration test (SMPT)

- It is the test to evaluate the capacity of the spermatozoa to travel through cervical mucus in female genitalia. Because of the rheological properties of mucus, there are many hindrances or resistances in the path of spermatozoa. Sperm cell travels through mucus by adapting the channel with the least resistance. If encounters any resistance, it deflects parallel and moves forward in the female genitalia. The sperm penetration through estrual mucus depends mainly on its sodium and protein contents.

- The capacity of spermatozoa to travel up to more distance through mucus (estrual discharge) is indicative of better seminal quality.**Vanguard spermatozoon** is the spermatozoom that travels the farthest distance in mucus.
- General procedure:
  - Collection of estrous mucus from female at standing heat (may be stored at -20°C)
  - Mucus is filled in a non-heparinized glass capillary tube (Length 80mm; Diameter 0.8 mm) using a fine needle
  - One end of mucus filled capillary tube is sealed with wax
  - The open end of the capillary tube is dipped vertically in a semen sample (0.5 -1 ml) taken in a microcentrifuge tube
  - The set up is incubated at 37°C for 30-60 min. The capillary tubes may be cooled quickly in crushed ice to give cold shock to prevent further sperm migration.
  - The capillary tube is placed on a graduated glass slide and the distance travelled by vanguard sperm is observed under a microscope.
- SPD (Sperm penetration distance) travel by vanguard sperm is indicative of semen quality. SDP:
  - ≥29 mm – Excellent semen quality
  - 20-29 mm – Good semen quality
  - ≤19 mm – Fair to poor semen quality
- Some authors reported that the fertile sperm travel 20 mm in 20 minutes through estrual mucus.
- To overcome the problem of collecting estrual mucus or declining its quality on storage, one can use 2% polyacrylamide gel instead of cervical mucus at estrus.

5.5. Post coital test

- Cervical mucus is collected from female genitalia after 3-4 hrs of insemination.
- One drop of collected mucus is observed under microscope.
- More motile sperms in collected mucus indicate the better semen quality.

- The presence of a single motile spermatozoon in collected mucus ensures that the cervical mucus may not be a possible reason for infertility.

6. DNA integrity

- Sperm with damaged DNA of various degrees may have normal membrane integrity, normal motility pattern and even normal fertilization with oocyte. Most probably there is abnormal embryonic DNA expression which is responsible for infertility. Poor chromatin quality, defective chromatin packaging or poor decondensation during ova penetration leads to infertility. DNA in sperm is arranged with protamine (a basic protein) in condensed or well packed state. Decondensation occurs because of the separation of protamine from DNA when sperm enter the ova. Abnormal DNA may be responsible for abnormal decondensation in sperm cell which leads to infertility.
- Sperm DNA abnormalities can be diagnosed by flow cytometric based sperm chromatin structur eassay (SCSA), Comet assay, TUNEL assay, cytochemical assays and transmission electron microscopy (TEM) etc.
- DNA condensation (well packed DNA) in sperm can be investigated with TEM, Aniline blue staining, Chromomycin A3 and TUNEL assay.
- Abnormal DNA decondensation ability can be investigated with Giemsa staining. Liberated DNA from sperm cell (Using SDS) is induced to be decondensed (By removing Zn by using chelating agent EDTA). After staining with Giemsa stain, sperm cells are observed under 100X bright field microscope. Swollen head sperms (indicates decondensed) are considered as of good quality pertaining to fertility.
- DNA fragmentation (Damage to strands of DNA) can be investigated with
    1. Acridine orange (Green fluorescence for double stranded normal DNA and orange/red for single stranded or denatured DNA)
    2. DAPI (Non fluorescence complex for double stranded DNA)
    3. SCSA; Sperms with abnormal DNA integrity are more prone to acid/ heat treatment which leads to denaturation of DNA (ssDNA). Acid or heat treated sperm (having denatured or ssDNA) are stained with acridine orange and evaluated with flow cytometry (quantification); Red/orange fluorescence for denatured DNA and green fluorescence for intact double DNA of spermatozoa is observed. DNA fragmentation index (DFI) should be <15% for excellent quality of

semen. A semen sample having DFI >29-30% is considered as poor quality.

4. Comet assay (single cell gel electrophoresis assay); DNA is liberated from sperm cell by using lysis buffer or detergent like SDS. Damaged DNA migrates on when subjected to electrophoresis, creating a comet like image which can be observed by staining with SYBR 14. More number of cells creating comet due to damaged DNA indicates poor seminal quality.

7. Acrosomal integrity

- After capacitation, spermatozoa undergo acrosomal reaction i.e. fusion and vesiculization of the outer acrosomal membrane with plasma membrane culminate in release of acrosomal content e.g. acrosin etc. Acrosome reaction makes sperm penetrate ova through zona pellucida.
- Sperms undergoing capacitation and acrosome reaction before reaching female genitalia indicate the poor fertility. It may happen due to cryocapacitation or pre-acrosome reaction during freezing due to egg yolk (presence of progesterone and Ca^{2+}) and many other factors.
- Acrosome reacted sperm can be assessed with Giemsa stain, chlortetracycline (CTC), Fluorescent stains (FITC-PSA/PI or FITC-PNA/PI) and triple stain (trypan blue/Bismark brown/rose Bengal or PE-PNA/SYBR/PI) etc.
- Giemsa staining : It stains the sperm acrosome which can be visualized as purple coloured stain with bright field microscope. Intact acrosome is observed as uniformly stained while damaged acrosome and acrosome reacted sperms are stained unevenly (rough) and unstained respectively.
- CTC staining : CTC is chlortetracycline which is a fluorescent antibiotic which has the affinity for Ca^{2+} and Mg^{2+} ions in sperm cells. It bounds to these ions in sperm plasma membrane and gives fluorescence.
  - Whole sperm head fluorescence - Uncapacitated sperm (F-pattern).
  - Only preacrosomal region fluorescence - Capacitated sperm (B-pattern).
  - No/dull fluorescence or only of the equatorial plate - AR reacted (AR-pattern).
- FITC-PSA/PI or FITC-PNA/PI – PSA and PNA are membrane impermeable lectins (carbohydrate binding agglutinin proteins) isolated from *Pisum sativum* and *Arachis hypogia*, respectively. These are

tagged with fluorescent dye FITC (fluorescein isothiocyanate). These agglutinins bind with galactose and mannose carbohydrate molecules on sperm cell acrosome and give fluorescence due to FITC.

- Uniform green fluorescence – Sperm cells with intact acrosome
- Uneven green fluorescence – Sperm cells with variably damaged acrosome
- No green fluorescence – Acrosome reacted sperm cells
- Red/Orange fluorescence – Dead sperm cells

- CTC can differentiate between capacitated and non capacitated sperm cells while FITC-PSA or FITC-PNA cannot differentiate between capacitated and non capacitated sperm cells. PI can differentiate between live and dead sperm cells.
- FITC-PSA has some affinity for egg yolk i.e. egg yolk molecules can be observed as sperm cells with intact acrosome. So it is better to use triple stain like PE-PNA/PI as it has no affinity for egg yolk.
- SYBR-14 and PI are DNA specific stain. Triple stain PE-PNA/SYBR14/PI is used to assess the acrosomal integrity of the spermatozoa. Orange fluorescence indicates the acrosome reacted sperms while intact acrosome sperm cells don't give orange fluorescence. Green and red fluorescence indicate the live and dead spermatozoa.

8. Mitochondrial activity

- Sperm mitochondria (50-70 per cell) are a producer of ATP as an energy source which is responsible for sperm motility and other activities.
- Rhodamine 123 is a cationic fluorescent molecule which gives fluorescence only In case of functional mitochondria. It is accumulated in active mitochondria and washed out from the cell with lost membrane potential.
- JC-1 and MTT can also be used for mitochondrial activity assessment in sperm cells. Mitochondria having high membrane potential, JC-1 enter the mitochondrial matrix where it accumulates and gives red fluorescence. In having mitochondria with low membrane potential, JC-1 cannot accumulate within the mitochondria and remains in the cytoplasm giving green fluorescence.
- A triple stain like FITC-PSA/PI/JC-1 is also used for the assessment of mitochondrial activity.

9. Other tests and techniques

9.1. Antisperm antibodies (ASA) test

- Some antisperm antibodies are present in the serum, seminal plasma of male and cervical mucus of female. These antibodies show an immune reaction against sperm cells and deteriorate the sperm quality.
- Primarily, IgG and IgA are considered as antisperm antibodies. IgA are abundant in seminal plasma while IgG in serum.
- Their detection can be done with Sperm Mar assay, Immune bead test, Tray agglutination test and Immobilization test.

9.2. Swim-Up test (sperm migration test)

High motile spermatozoa have the ability to get separated from other cells or debris in seminal plasma and swim up actively into the overlaid supernatant medium. High motile spermatozoa can be achieved by centrifuging the semen in a percoll gradient, by glass wool filtration or by Sephadex filtration.

9.3. Various **epididymal proteins** like osteopontin, lipocalin type prostaglandin D-synthetase (L-PGDS) etc are indicators of good semen quality.

9.4. Computer assisted sperm analysis (CASA)

In CASA large no. of sperm cells are evaluated for various parameters like concentration, viability and motility within a very short time. It is a more effective, precise and reliable tool for semen quality assessment. Results are highly predictable in terms of fertility potential. Motility in terms of various parameters like progressive, slow, linearity of sperm movement, the amplitude of the lateral head displacement, the beat cross frequency etc can be analyzed with the help of CASA.

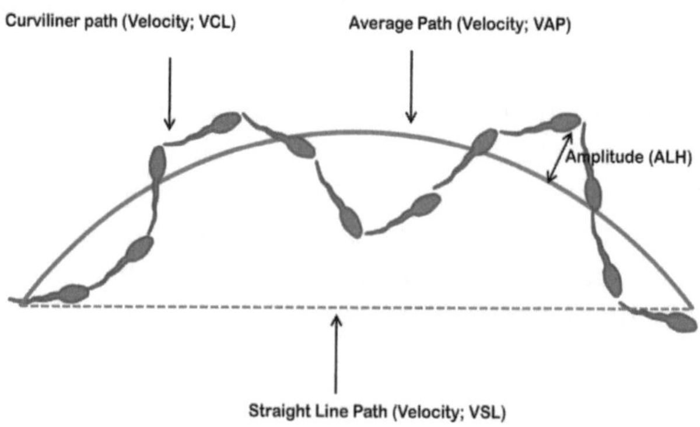

Fig. Sperm motility and velocity parameters measured by CASA systems.

Parameter	Unit	Description
M (Motility)	%	Cells that are motile with more than a minimum specified speed
PM (Progressive motility)	%	The population of cells that are moving in Forward direction
VCL (Curvilinear velocity)	μm/s	Average velocity of a sperm along its actual curvilinear track
VAP (Average path velocity)	μm/s	Average velocity of sperm along its average curvilinear path
VSL (Straight line velocity)	μm/s	Average velocity of a sperm head along the straight line between sperm displacement (Start to endpoint)
ALH (Amplitude of Lateral Head Displacement)	μm	Average deviation of the sperm head from average curvilinear path
BCF (Beat Cross Frequency)	Hz	The frequency with which the head moves back and forth in its track across the cell path.
STR (Straightness)	%	Linearity of the average path (VSL/VAP)
LIN (Linearity)	%	Linearity of the curvilinear path (VSL/VCL)

9.5. Flow cytometry

Fluorochrome bound cells, suspended in a fluid stream, are moving with high velocity. The emitted light is responded by photomultiplier tubes and converted into electronic signals which are interpreted by cytometry software. Thousands of cells are analyzed within seconds for multiple parameters. Structural and functional characteristics of spermatozoa, including plasma membrane integrity, acrosome integrity, mitochondrial activity, DNA structure and morphological abnormalities can be analyzed by flow cytometry.

9.6. Scanning or Transmission electron microscopy (SEM/TEM) is used for observing sperm morphology and mitochondrial structures)

9.7. Sephadex filtration method (motile and non motile sperms, dead, damaged, uncapacitated sperms etc)

9.8. SQA (semen quality analyzer; motility, concentration, morphology etc)

12

Sperm Structure

Akhil Patel and Vikas Sachan

Semen

- Semen, a fluid released from the male reproductive system during ejaculation, is made up of a liquid phase called seminal plasma and a cellular component called spermatozoa.
- The term **"spermatocrit"** refers to the percentage of spermatozoa(v/v) in semen, which varies from less than 2% in pigs to more than 30% in sheep. It is about 10% in bull.

  Semen= Spermatozoa + seminal plasma

Spermatozoon

- In 1677, the scientist Van Leeuwenhoek first discovered the spermatozoa and called them **"animalacula"**.
- Spermatozoa or sperm cells or male gametes, develop within the wall of the seminiferous tubules as a result of a complex process known as spermatogenesis.
- The spermatozoon is a highly specialized cell that performs the sole function of fertilizing an oocyte.
- Spermatozoa have distinct morphological characteristics that are essential for cell function. Although spermatozoa vary in size and shape between species, the basic morphologic features are constant. Deviation from specific morphologic structural correlates with male sub- and infertility. Sperm configuration can be surveyed well with light microscopy (LM). Electron microscopy (EM) delivers a display of inner and outer structure of spermatozoa more finely.
- The shape of sperm head is generally oval but in rat it is falciform/sickle/ hook shaped & in cock is spike shaped.

- The head and the tails are the two main, mutually dependent components of spermatozoa. Some authors consider the neck portion separately. The apparatus required for energy production and movement is located in the tail, whereas the sperm head contains the components required for fertilization and paternal DNA.

Head

- The shape of the sperm head is species specific and usually flattened dorso-ventrally. It is pin-shaped in chicken while hook shaped in rat. Head of cattle bull spermatozoon is longer while buffalo bull spermatozoo, it is wider.
- The sperm head may be divided into two sections: the anterior acrosome (glea capitis) and the posterior post-acrosomal region. The acrosome and nucleus are surrounded by the plasma membrane.
- The nucleus lies beneath the anterior acrosome and extends to the base of the head. The nucleus is the largest part of the sperm head and contains highly compacted DNA by sperm-specific proteins protamines. The cell is non-dividing and transcriptionally inactive due to its compaction of DNA by protamines. The nucleus is encircled by the post-acrosomal sheet, a cytoskeletal complex of perinuclear theca.
- The acrosome is a unique sperm organelle that forms during the early stages of spermatogenesis from the Golgi apparatus. The acrosome is a thin double-layered membranous sac that covers roughly two third of the apical area of the sperm nucleus and head in a cap-like fashion. The acrosomal membrane directly beneath the plasma membrane is known as the outer acrosomal membrane, whereas the acrosomal membrane above the nucleus is known as the inner acrosomal membrane. The acrosome's posterior part is known as the nuclear ring or the equatorial segment. Because the acrosome emerges from the Golgi apparatus during spermatogenesis, it has been referred to as a specialized lysosome.
- Acrosome contains hyaluronidase, proacrosin, acrosin, esterases, neuraminidase, acid phosphatase, phospholipases, aryl phosphatase, β-N-acetylglucosaminidase, arylamidase, collagenase and corona penetrating enzyme (CPE). Hyaluronidase plays a role in either the dispersal or the digestion of the cumulus oophorus extracellular matrix of the oocyte. CPE plays a role in penetrating through the corona radiata of the oocyte. Acrosin is present in the acrosome as proacrosin; is converted into acrosin. Acrosin is a trypsin-like enzyme that functions to allow penetration of spermatozoa through the zona pellucida during

fertilization. Neuraminidase also evidently enhances the passage of spermatozoa through the zona pellucida and the plasma membrane of the oocyte. Actually plasma membrane at the equatorial segment is responsible for the contact with the oocyte membrane during fertilization.

Neck

- The neck is a short linking segment between the sperm head and flagellum i.e. connecting head to middle piece.
- The neck is approximately 1 μm and it is attached anteriorly to the basal plate and posteriorly to the outer dense fibres (ODF) of the flagellum. It is composed of segmented columns and a dense fibrous structure, the "capitulum". Next to the capitulum, there is the proximal centriole whereas the distal centriole is absent in mature spermatozoa.

Tail

- The tail is the longest part of the sperm and is divided into the middle piece, principal piece, and endpiece.
- A common cell membrane surrounds each of these parts. The axoneme, mitochondrial sheath, outer dense fibres (**ODF**), and fibrous sheath are the key structural components of the mammalian sperm tail.
- The middle piece is the portion of the tail between the neck and the **annulus (Jensen's ring)**. The annulus or Jensen's ring is a structural component of the mammalian sperm tail that joins the middle piece to the principal piece.
- In the midpiece, there is a centrally located axoneme made up of a central pair of singlet microtubules surrounded by nine evenly spaced microtubule doublets joined with Nexin links i.e. (9X2+2). The axoneme ((9X2+2)) is encircled with nine outer dense/coarse fibers (ODFs). This central skeleton, axoneme surrounded by ODFs, runs from the neck to principal piece (except the end piece).
- The outer dense fibre-axoneme complex in the middle piece segment is covered by a helically (spiral) wrapped mitochondrial sheath. The mitochondrial sheath (layer of mitochondria) is made up of three to four spirals of nearly 70 mitochondrias. The number of mitochondria and spirals may be more. This mitochondrial arrangement in the middle piece is surrounded by a thin layer of cytoplasm called **"manchette"** which is ultimately surrounded by the plasma membrane.

- The principal piece (from the annulus to the terminal piece) is the longest portion of the sperm flagellum. The diameter of the tail in the principal piece is shortened due to the termination of mitochondria in the middle piece. In the principal piece, the outer dense fibre-axoneme complex is surrounded by a fibrous sheath which is encircled with the plasma membrane. ODFs-axoneme complex with fibrous sheath taper to almost absent at the principal piece termination point.
- The end piece is the region beyond the distal end of the principal piece. Here fibrous sheath along with ODfs is absent i.e. axoneme is surrounded with only plasma membrane.

S.No.	Species	Length of spermatozoa (μm)	Length of the sperm head (μm)
1.	Bull	70-90	9-9.5
2.	Stallion	55-65	5
3.	Ram and Buck	60-80	8-8.5
4.	Boar	50-60	8.2-9.0
5.	Dog	55-65	4.5-7.0
6.	Tom	55-65	5.5-6
7.	Cock	90-100	10-20
8.	Rat	190-195	15-15.5
9.	Man	50-65	4-6

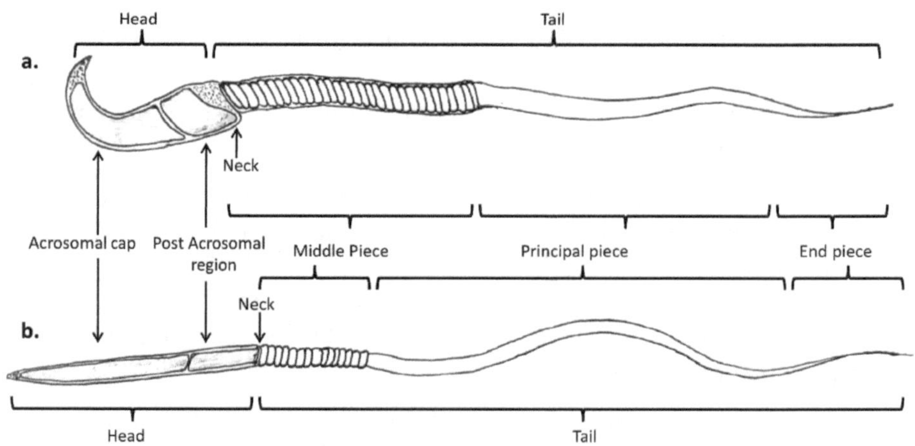

SPERMATOZOON STRUCTURE : a. Rat b. Chicken

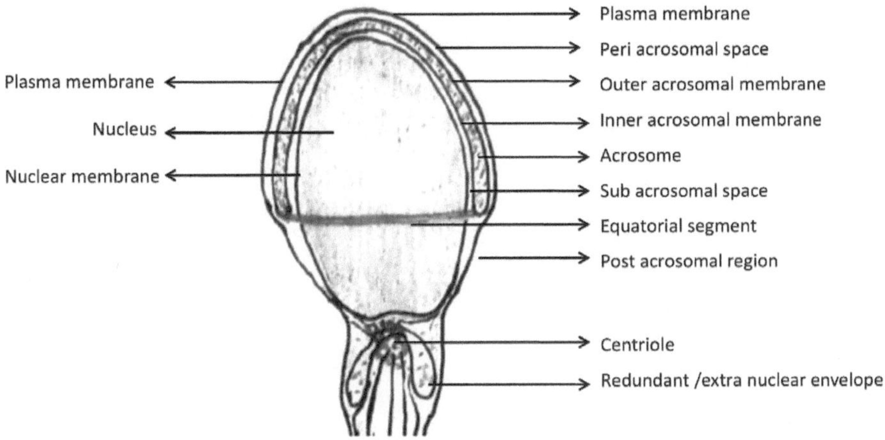

SPERM HEAD

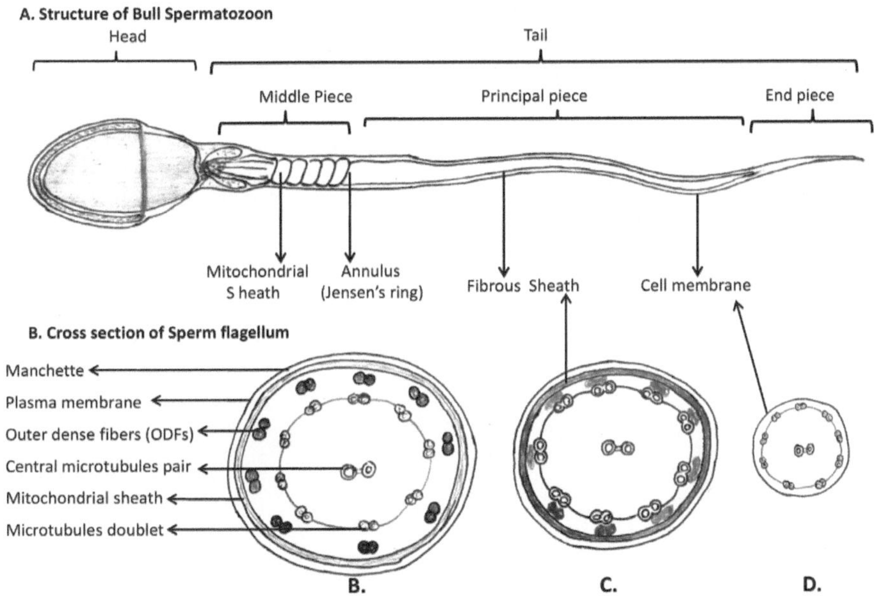

Spermatozoon structure: A. Structure of a bull spermatozoon, **B.** Cross section of sperm flagellum; I. Middle piece, II. Principal piece, III. At end piece

Spermatozoon structure: A. Structure of a bull spermatozoon, **B.** Cross section of sperm flagellum; **I.** Middle peice, **II.** Principal piece, **III.** At end piece

13

Sperm Abnormalities

Vikas Sachan

- Sperm abnormalities may be genetical or morphological. They are directly related to the male infertility.
- The wet mount method and phase contrast microscopy facilitate sperm visualization for abnormalities. Semen stained with vital stain can also be used for evaluating sperm abnormalities. Sperm morphology or related abnormalities can be evaluated by using eosin-nigrosine stain, carbolfuschin, aniline blue-eosine etc.
- Generally, for head and tail abnormalities rose Bengal and William's stain can be used while for acrosome and head abnormalities Giemsa stain can be used.

Categorization of Sperm Abnormalities

1. On the basis of site of abnormality on the spermatozoa

- *Head abnormalities* e.g. microcephalic head, macrocephalic/giant head, narrow or short sperm head, pyriform head, detached glea capitis (acrosomal cap), detached head and double head etc.
- *Midpiece abnormalities* e.g. Distal midpiece reflex, double midpiece, proximal and distal cytoplasmic droplets and cork screw defects etc.
- *Tail abnormalities* e.g. broken tail, bent tail, double tail, coiled tail, dag defect etc.

2. On the basis of the site of origin in male genitalia

- *Primary sperm abnormalities*: these abnormalities are due to a defect in spermatogenesis i.e. abnormality originates at the germinal epithelium in seminiferous tubule. e.g. Knobbed sperm, diadem sperm defect, mid piece defect (swollen mid piece, abnormal mitochondrial sheath), microcephalic head, macrocephalic/giant head, narrow or short sperm head, pyriform head, proximal protoplasmic droplet (PCD), double head/midpiece/tail sperm, dag defect, abaxial tail sperm defect, tail stump defect etc.

- *Secondary sperm abnormalities*: These abnormalities originate after spermatogenesis and spermiation i.e. during the travel in male genitalia. e.g. proximal protoplasmic droplet, distal cytoplasmic droplet (DCD), detached glea capitis (acrosomal cap), detached head, broken neck, broken/bent tail etc
- *Tertiary sperm abnormalities*: These abnormalities originate after ejaculation i.e. during processing or handling the semen due to mechanical (rough handling), osmotic (change in pH or osmolarity) or thermal (change in temperature) shock.e.g. broken tail, detached head, bent tail, coiled tail etc
    - Abaxial tail sperm defect may be normal in stallion and boar. PCD and double head are also commonly found in stallion semen.

3. On the basis of the effect on fertility

a. *Major sperm defects*: Knobbed sperm, diadem sperm defect, PCD, Dag defect, double head/midpiece/tail, abnormal head, tail stump, distal midpiece reflex etc

b. *Minor sperm defects*: small/giant head, abaxial tail sperm defect, detached acrosomal cap, DCD, bent tail etc.

4. On the basis of compensating infertility

a. *Compensable sperm defects*: In these type of sperm defects, spermatozoa are not able to travel through the uterine tube up to ova or not able to penetrate ova. Fertility can be achieved by doing insemination with an increased volume of such semen samples e.g. knobbed sperm, loose/ detached head, dag defect, distal midpiece reflex, tail stump, abnormal mitochondrial sheath etc.

b. *Non-compensable sperm defects*: In these type of sperm defects, spermatozoa are able to travel up to ova and penetrate ova but there is no cleavage or non-viable embryo formed. Fertility cannot be achieved even by increasing inseminating seminal volume. e.g. PCD, pyriform head, nuclear defects, micro/macrocephalic sperm head etc.

Some Important Sperm abnormalities

1. Diadem Sperm defect

- There is invagination or inpouching in the nuclear material of spermatozoa. Invaginations may be seen as beaded or pearl like structures (resembling beads in a necklace) at the equatorial segment of spermatozoa is seen. Pouches may be single or more and maybe somewhere else on the sperm head.
- Eosin-nigrosin stain and fuelgen stain can be used to detect such defects.
- It is a primary (defect of spermatogenesis), major and uncompensable sperm defect.
- Sperm with this defect may have normal motility.

2. Knobbed sperm defect

- In this defect, there is accentrically placed thickening of acrosome i.e. enlarged acrosomal cap in spermatozoa.
- It is a genetic sperm defect (inherited, autosomal recessive sex-linked defect) or generated due to defective spermatogenesis. It is commonly found in bull, boar and dog.
- Eosin-nigrosin stain and fast green stain can be used to detect it.
- Sperm motility and concentration may be normal but spermatozoa are unable to penetrate ova.
- It is primary, major and uncompensable sperm defect.

3. Protoplasmic droplet sperm defect

- During the process of spermiation, spermatozoa are released with a cytoplasmic droplet at the neck region along with some residual bodies. Residual bodies are phagocytized by Sertoli cells and thus a cytoplasmic droplet remains at the neck of spermatozoa. During the maturational process this proximal cytoplasmic droplet migrate to distal mid piece region i.e. converts into distal cytoplasmic droplet in caput epididymis. Distal cytoplasmic droplet is further lost during traverse of germ cell in male genitalia before ejaculate.
- The persistence of protoplasmic droplet at the neck of spermatozoa (proximal protoplasmic droplet) or at the distal mid piece region (distal protoplasmic droplet) is considered as sperm defect and indicates the immature spermatozoa. Such sperm defects are common in boar semen

- PCD is primary (defective spermatogenesis) or secondary (defective epididymal maturation), major and non compensable sperm defect while DCD is considered as secondary and minor sperm defect.

4. Pseudo-droplet sperm defect

- There is round or irregular unsymmetrical thickening at the region of mid piece. This defect should be differentiated with a protoplasmic droplet (spherical).
- Spermatozoa having droplet defect has reduced motility.

5. Dag defect

- There is strong coiling of the main piece or tail over the mid piece of the spermatozoa (may be appeared as a short tail).
- It is due to abnormality in axial filaments and mitochondrial sheath. Such abnormalities occur either at the level of testis or in epididymis.
- It may be either hereditary or due to high Zn in the diet.

6. Corkscrew sperm defect

- Irregular helical arrangement of mitochondria with irregular distribution of mitochondrial sheath is found in mid piece of spermatozoa giving an appearance like cork-screw.
- This defect may be inherited, due to testicular degeneration, or due to some environmental factors. Spermatozoa with cork-screw defect are immotile or have abnormal motility.

7. Tail stump sperm defect

- Defected spermatozoa have a small tail portion (2-3μm) with little amount of flagellar material.

8. Distal midpiece reflex

- The Bending of midpiece of spermatozoon engaging a protoplasmic droplet at the point of bending.

9. Decapitated or disintegrated sperm defect

- The head is detached from tail due to some structural defect at neck region i.e. head and tail are separated. It is due to defective spermatogenesis.

10. Acrosomal defect

- The acrosomal cap (glea capitis) of the spermatozoon is damaged with various degrees. Sometime complete evulsion of the acrosomal cap is seen.

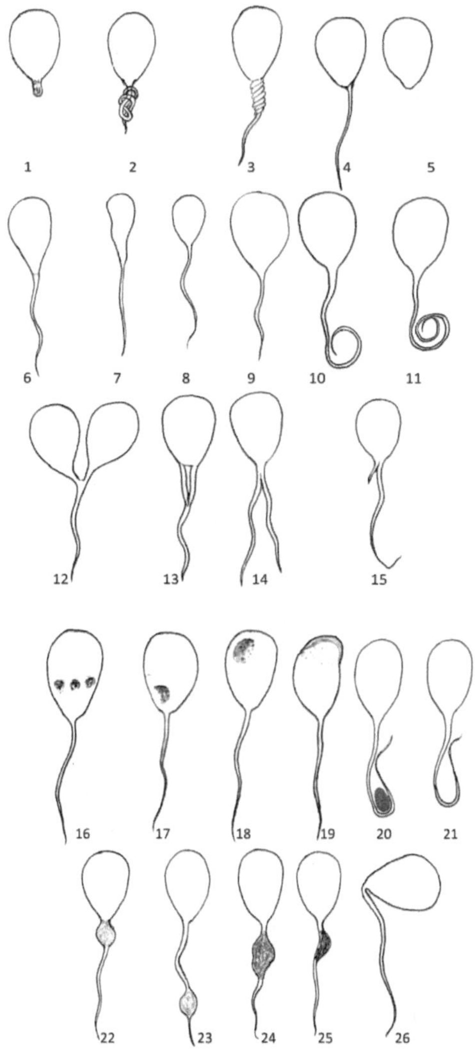

Various Sperm Abnormalities

1. Stump tail 2. Dag defect 3. Cork screw sperm 4. Abaxial tailed sperm 5. Detached head 6. Pyriform head 7. Abnormal head 8. Microcephalic head 9. Giant head 10-11. Coiled tail 12. Double head 13. Double midpiece 14. Double tail 15. Accessory tail 16-18. Diadem defect 19. Knpbbed sperm 20. Distal midpiece reflex 21. Looped tail 22. PCD 23. DCD 24-25. Pseudo droplet defect 26. Broken head

14

Semen Cryopreservation

Atul Saxena and Vikas Sachan

- **Semen cryopreservation** – It is the process of preserving the semen in a suitable medium at an ultra-low temperature which reduces spermatozoa metabolism with the least production of metabolic waste products of sperm cells to prolong their life span.
- Every fall of 10^0C temperature decreases the metabolic rate of spermatozoa to half.

Principle of various components involved in cryopreservation

1. Low temperature

- Reduce metabolism (negligible) and least metabolic waste product
- Decrease solute solubility leads to increased viscosity and physical resistance
- Reduce motility
- Prolong life span

2. Protection from cold shock and cryoinjury

- When sperm cells are subjected to a reduction in temperature during the freezing process, there are:
  - Sperm membrane damages and lead to leakage of intracellular components like enzymes, ATP, K^+, cholesterol, lipoproteins etc.
  - Irreversible abnormality in motility pattern (bending of sperm tail)
- Cold shock protecting agents like lecithin, lipoproteins, phospholipids etc (present in milk and egg yolk) are present in the extender which protects sperm cells at 3-5^0C i.e. chilling temperature or refrigeration temperature.

- Cryoprotectants e.g. DMSO, glycerol and ethylene glycol etc. protect sperm cells at ultra low temperature (freezing temperature) by reducing the formation of ice crystals.

3. Source of energy

- The Source of energy is required for the aerobic and anaerobic metabolism of sperm cells. Breakdown of sugar like glucose, fructose, mannose, arabinose etc generates ATP to provide energy for sperm activity
- Probably, the source of energy is utilized by sperm cells for their survival up to freezing and after thawing.

4. Osmotic pressure

- Semen extension media is isotonic to seminal plasma (280-300 mOsm/l). Sperm cell can tolerate moderate change in osmolarity.
- Hypo-osmotic and hyper-osmotic media causes swelling/bursting (due to endosmosis) and shrinking (due to exosmosis) of sperm cells respectively. It is called as **solution effect**. Non-isotonic extension of semen leads to tail bending, circling movement or death of the sperm.

5. Buffering capacity

- Spermatozoa produce lactic acid due to the breakdown of sugar (metabolic activity) which changes the pH of semen.
- While sperm has a narrow range of tolerance to change in pH, buffer (egg yolk, milk, citrate buffer, phosphate buffer, TRIS, TES, HEPES etc) are used to maintain the seminal pH.

6. Dilution effect

- When semen is diluted (to achieve 20 million sperms in a straw dose of 0.25 ml), there is a reduction in viability & motility of the sperm cells and thence fertility. It is due to the leaching of some intracellular as well as structural components from cell membrane along with the loss of protective seminal plasma components that leads to capacitation like changes during dilution. This detrimental effect is referred as **dilution effect.**
- Protective agents like lecithin and low density lipoproteins (LDL) in egg yolk are used in an extender. They prevent spermatozoa from dilution effect to maintain their fertilizing capacity.

7. Antimicrobial activity

- Entry of microbes in semen occurs during collection as well as processing. Their proliferation is supported due to growth media such as milk or egg yolk in the extender.
- These microbes deteriorate seminal quality by:
  - Compete with spermatozoa for the limited nutrition availability
  - Production of various toxins and ROS which are harmful to sperm cells
- Even if there are non-pathological microbes in semen or diluted semen, the antibiotic is added to reduce the competition for available nutrients. Aseptic precautions along with the incorporation of some antimicrobial agents like penicillin, streptomycin, polymyxinB etc. protect spermatozoa from the detrimental effects of various microbes.
- A combination of 1000 units of penicillin and 1000 µg of streptomycin is found to be sufficient for each ml of extender. Excess antibiotics in semen may be toxic for the spermatozoa without any extra beneficial effect. Gentamicin and amikacin may be used in equine semen dilution.
- Lincomycin and spectinomycin can be added for protection from mycoplasma and ureaplasma.
- Sulphanilamide crystallizes during freezing while procaine penicillin is toxic to sperms so these antibiotics are not used.

8. Protection from toxic materials

- Double or triple distilled water is used during diluter preparation to avoid any metal toxicity.
- Only analytical grade of reagents is preferred to avoid any type of toxicity.
- Glycerol is somewhat harmful to sperm cells but there is no better substitute till date.

9. Maintenance of gas phase

- Air space is maintained over the diluted semen in order to provide oxygen for sperm metabolism.
- In the presence of CO_2, Sperm activity is reduced which prolongs the life span.

Steps involved in the process of deep freezing

- Preparation of bull
- Preparation of extender and keep at 32^0-35^0C.
- Collection of semen with artificial vagina and keep at 32^0-35^0C
- Seminal evaluation
- Semen extension with a fraction of diluter without glycerol up to 1:1 (at 32^0-35^0C)
- Seminal evaluation and calculation of the final dilution rate
- Cooling down the diluted semen to 4^0C in 1 to 1.5 hrs
- Final dilution with a fraction of diluter with glycerol (at 3^0-5^0C)
- Equilibration of diluted semen (4^0C for 4-5 hrs)
- Printing and filling of semen straws (0.25 ml) with diluted semen and sealing (with polyvenyl alcohol powder (PVA) or machine sealing)
- Vapour freezing of cooled semen straws (manual or automatic freezing)
- Storing the frozen straws in goblets and canisters in liquid nitrogen cryocan
- Washing and sterilization of all the glassware for next day usage

- **Process of cryopreservation in a flow diagram**

Collection of fresh semen (Using an artificial vagina)

↓

Physico-morphological tests

Volume, Colour, Consistency, pH, Gross motility, Sperm concentration, Progressive motility

↓

Semen dilution with Egg yolk Tris extender (With Glycerolization)

↓

Seminal examination after dilution

(Progressive motility, Live percentage, Morphological abnormality, Acrosomal integrity, Membrane integrity (HOST) etc.)

↓

Straw filling and sealing

↓

Equilibration (4° C for 5 hr.)

↓

Seminal examination after equilibration
(Progressive motility, Live percentage, Morphological abnormality, Acrosomal integrity, Membrane integrity (HOST) etc.)

↓

Vapour freezing in programmable biological freezer

↓

Storing of straw in LN2 (-196° C)

↓

Thawing

↓

Post thaw seminal evaluation (after 24 hours of freezing)
(Progressive motility, Live percentage, Morphological abnormality, Acrosomal integrity, Membrane integrity (HOST) etc.)

15

Semen Extension

Vikas Sachan and Hanuman Prasad Yadav

Semen is needed to be extended with a suitable medium and cryopreserved without affecting its survivability and fertility for the purpose of inseminating female artificially.

Qualities of Ideal Extenders

- Semen extender is a nutrient media which provide energy to spermatozoa, protection from osmotic shock, thermal shock and various metabolic wastes for prolonging the life span of sperm cell with maintaining the acceptable fertilizing capacity. A semen extender should have qualities to protect spermatozoa during dilution and cryopreservation. These qualities may be :

  1. Preserving sperm cell integrity (membrane, cellular, mitochondrial and acrosomal integrity etc)
  2. Maintenance of osmotic pressure (isotonic; 280-300 mOsm/l) similar to seminal plasma
  3. Providing nutritive elements as energy source for sperm metabolism (aerobic & anaerobic metabolism)
  4. Having buffering capacity (to resist any change in pH due to sperm metabolism)
  5. Protecting from microbial contamination (having antimicrobial components)
  6. Protecting sperm cells from cold shock (Lecithin, lipoproteins, phospholipids etc) and cryoinjuries (cryoprotectant)
  7. Providing all essential elements for extending the survivability of sperm cells
  8. Clear visibility of diluted semen under microscope
  9. Easily available and economic

History Related to Semen Dilution

- Term extender was used in place of diluter by Foote and Bratton (1950) as it is the extension of the usefulness of semen up to minimum spermatozoa required for conception.
- Philip & Lardy were the first to use egg yolk as a cold shock protectant for spermatozoa.
- Egg yolk phosphate extender was discovered by Philip &Lardy (1940) and egg yolk citrate (sodium citrate) was discovered by Salisburry (1941).
- Glycerol was used 1st time in semen extender by Polge, Smith and parker. They were the first to freeze fowl and bull spermatozoa using glycerolized extender in 1949.
- Davis et al. (1963) concluded that Tris-buffered egg yolk-glycerol provided excellent protection for sperm either frozen or unfrozen. Now this extender is world widely accepted bull sperm and for other species.
- Antibiotic was first used in extender by Almquist in 1946. Cornell extender (CUE) was the first extender with antibiotic (penicillin, streptomycin and polymyxin B).

Semen Diluters

1. For semen preservation at room temperature (ambient temperature; 18^0-30^0C): It does not contain cryoprotectant (e.g. glycerol or DMSO) but little amount of cold shock protectant (e.g. egg yolk) may be added

a. **Coconut milk extender (CME)**
   - Coconut milk is collected, boiled, cooled and then filtered.
   - It contains 7% egg yolk and catalase.
   - It should be kept in a dark place with no air space in vials/bottles
   - Its pH value is 7.4
   - Semen preservation can be done for 3-4 days

b. **Millovanov's diluter**

c. **Carbonate phosphate**

d. **Caprogen:** N2 gassing is done in this extender.

e. **Illini variable temperature diluter (IVT)**

- Exposure to CO_2 bubblesfor 10 minutes (pH 6.3)
- It contains 10% egg yolk.
- Diluted semen can be preserved for 3 days

f. Cornell university extender (CUE)

- CO_2 is produced due to the reaction of citric acid and bicarbonate i.e. no need of exogenous CO_2
- CO_2 reduces metabolism and motility of spermatozoa but long exposure may be toxic.
- Diluted semen can be preserved for 8 days but fertility goes on reducing after 48 hours.

2. For semen preservation at refrigeration/chilling temperature (4^0-5^0C): It contains cold shock protectant (e.g. egg yolk) but no cryoprotectant (e.g. glycerol or DMSO).

a. Milk diluter

- It Contains homogenized or skimmed or reconstituted milk powder. Skimmed milk is preferred as it has less or no fat globules so microscopic visibility is clearer.
- There is a spermicidal factor in milk called "**lactanin**". It is inactivated with heat treatment (90^0-95^0C for 5 minutes) or chemical treatment (cystine or thioglycolic acid convert toxic lactanin to non toxic β-lactoglobulin).
- Diluted semen can be preserved for 3 days.

b. Citric acid whey diluter (CAW)

c. Egg yolk-phospahte diluter

- It is the 1st semen extender discovered (by Phillip and Lardy, 1939)
- Preservation of cattle bull semen can be done for 3-4 days (2 days buffalo semen).

d. Egg yolk citrate diluter (discovered by Salisbury)

- It contains 20% egg yolk for bull semen (7% for ram and 2% for buck semen).
- Preservation of cattle bull semen can be done for 3-4days (2 days buffalo semen)

e. **Kampschmidt's diluter**

f. **D_2 diluter**

3. For semen preservation at freezing/ultralow/subzero temperature (-79⁰ to -196⁰C): It contains cold shock protectant (e.g. egg yolk) as well as cryoprotectant (e.g. glycerol or DMSO). Semen can be preserved for the indefinite time provided that the cold chain is maintained.

a. **Glycerolate egg yolk citrate diluter**

b. **Glycerolated egg yolk TRIS diluter**

c. **Glycerolated milk diluter**

Different Semen Diluters For Different Species

S.No.	Species	Diluter
1.	Equine	Non-fat dried skimmed milk
		Mare's milk
		Kenney extender
		Cream gel extender
		Gelatin cream extender
2.	Boar	Kiev diluter
		Beltsville diluter (BF5)
		IVT
		Zorlesco
3.	Dog	Tris-fructose-citric acid
		Citrate carbonate egg yolk
		Citrate egg yolk
		Skimmed milk

Most of the diluters for equine and boar semen extension contain gelatin and EDTA respectively. Gelatin acts as a membrane stabilizer and EDTA acts as chelating agent (block calcium mediated capacitation and acrosome reaction).

Collection of Egg Yolk

- It should be done in laminar flow under sterilized condition
  - Take "A" grade fresh egg
  - Clean it with 70% alcohol swab
  - Crack and open the shell at the tapered end of the egg
  - Widen the opening gently with the help of foreceps
  - Discard out the albumin gently (**Lysozyme is found in albumin which is toxic to sperm cells**)

- Take the egg yolk ball (egg yolk enclosed in vitelline membrane) on filter paper
- The egg yolk ball is rolled over filter paper gently to soak any trace of albumin
- Puncture the vitelline membrane with sterilized forceps
- Collect the egg yolk in a sterilized graduated cylinder or beaker (egg yolk membrane should remain on filter paper)
- Egg yolk should be churned/beaten vigorously with a sterilized glass rod for few minutes in order to break large fat globules into micro-fat globules for better visualization under a microscope. Also, this reduces the chances of sperm head adhesion with the globule. The use of sodium citrate (as in egg yolk citrate buffer) makes fat globules dispersed in the egg yolk.

Preparation of Semen Extender

- Glycerolated TRIS egg yolk (EYGT) extender is used most commonly.
- Composition of EYGT:

S.No.	Component		Amount
1.	Buffer	TRIS (Hydroxymethyl aminomethane)	3.0284 gm
		Citric acid monohydrate	1.675 gm
2.	Energy source	Fructose anhydrous	1.250 gm
3.	Antibiotic	Penicillin G sodium	1 lakh unit
		Dihydro streptomycin sulfate	100 mg
4.	Cold shock protectant	Egg yolk	20 ml
5.	Cryopreotectant	Glycerol	7 ml
6.	Distilled water		Up to 100 ml

- Buffer can be autoclaved (5 psi for 10 minutes) and stored at 4^0C.
- pH should be 6.7 to 6.9 (pH< 5.5 is spermicidal)
- Osmolarity should be 280 to 300 mOsm/l.
- Antibiotic and fructose should be added before the dilution process.
- Egg yolk should also be mixed immediately before semen dilution as it may promote microbial growth as well as reduce the potential of antibiotics.
- To protect spermatozoa from oxidative damage (lipid peroxidation of the membrane), various antioxidants like glutathione peroxidase, catalase, cysteine, curcumin, Vit E etc can be added in extender. Membrane

stabilizers like chloroquine diphospahte and cholpromazine can also be used.

- Mixing of components can be done with a sterilized glass rod and magnetic stirrer.
- Preparation of the final extender should be done just before the semen dilution. The final prepared extender should not be preserved.

Dilution Rate

The objective of semen dilution is achieving a final concentration of 80 million sperms / ml, i.e. 20 million spermatozoa should be there in a French mini straw of 0.25 ml.

calculation

Volume of a semen ejaculate = X ml

Concentration of spermatozoa in fresh semen = Y million sperms / ml

Progressive motility in ejaculate sample = Z %

Effective concentration of progressively motile spermatozoa (A) = Y (Z / 100)

Final concentration to be achieved = 80 million sperms / ml

Dilution rate = 1 : A/80

Example

Volume of a semen ejaculate = 5 ml

Concentration of spermatozoa in fresh semen = 1600 million sperms / ml

Progressive motility in ejaculate sample = 80 %

Effective concentration of progressively motile spermatozoa (A) = 1600 (80 / 100)

A = 1280 million sperms / ml

Final concentration to be achieved = 80 million sperms / ml

Dilution rate = 1 : 1280/80 = 1 : 14

- i.e. 1ml of fresh semen should be diluted with 13 ml of diluter to achieve a concentration of 80 million spermatozoa / ml
- i.e. 5 ml of fresh semen will become 65 ml (5 X 13 = 65) extended semen after dilution.

- i.e. 280 (65 X 4 = 280) French mini straws (0.25 ml) can be prepared with 5 ml of fresh semen as four straws can be prepared from 1 ml diluted semen.

Semen Dilution Process

- **For chilled semen (refrigeration temperature):**
    - Extension should be done at 35^0-37^{0}C.
    - Semen dilution should be done up to 1:1 and allowed to mix properly for 20 minutes.
    - Check the various seminal attributes e.g. viability and progressive motility etc.
    - Final dilution should be done in 3-4 fractions at interval of 10-15 minutes to achieve a final concentration up to 10 to 20 million sperms/ml.
    - Diluted semen is filled in a dark vial with no air space and stored at 4^0-6^{0}C.
- **For preservation at ultralow temperature**
    - The extender is prepared into two fractions:
        - Fraction A without glycerol (0% glycerol) at 35^{0}C-37^{0}C
        - Fraction B with glycerol (14% glycerol) at 4^{0}C
    - Fraction A and B may be made with 3% and 11% glycerol.
    - Fraction A of extender is added to semen (at 37^{0}C) till 1:1 dilution slowly
    - Allow to mix for 15-20 minutes and check the various seminal attributes meanwhile.
    - Further dilution is done with fraction A till half dilution (according to dilution rate).
    - Reduce the temperature of this diluted semen to 4^{0}C by keeping it in cold handling cabinet (every reduction of 10^{0}C leads to decrease in the sperm metabolism approx to half).
    - Add fraction B into this low temperature diluted semen gradually up to final dilution to achieve 80 to 100 million sperms /ml.
    - Keep this finally diluted semen for equilibration.
    - The addition of glycerolated extender at 5^{0}C to semen is called **glycerolization**.

- It can be a added as a split method (addition with two equal extender fractions having 3% and 11% glycerol) or as a straight method (addition of extender having 7% glycerol).
- It should be done slowly (total time should be 30-45 minutes) at 5°C. Fast glycerolization at higher temperature may lead to damage to spermatozoa as well as leads to poor post thaw motility (PTM).
- 7% Glycerol is used in semen extender for cattle or buffalo bull semen. It is 2% in buck and boar semen extender.

Precautions For Semen Dilution

- Dilution should be done with freshly prepared semen diluter.
- The temperature of semen and semen diluter should be same.
- A little amount of diluter should be added at the start of dilution slowly to semen (especially till 1:1 dilution). Semen should never be added to diluter (it may cause osmotic shock to sperm cells).
- Initial progressive motility and concentration are estimated and then dilution rate is calculated. Final dilution should be done accordingly to achieve 20 million sperm cells per 0.25 ml straw.

Egg Yolk As A Cold Shock Protectant

- Cold shock occurs while lowering the seminal temperature from 37°C to 5°C during the freezing process.
- Cold shock affects sperm cell by:
    - Damage to the cell membrane
    - Leakage of intracellular K^+, enzymes, lipid, lipoprotein, cholesterol and ATP etc.
    - ***Phase separation*** i.e. conversion of liquid phospholipid of membrane into gel phase. This causes irreversible clustering of membrane protein (loss of function).
    - Loss of ATPase activity of membrane
    - Membrane permeability for Ca^{2+} increases
    - Alterations in capacitation process
- Sperms can be protected from cold shock by adding egg yolk, milk (skimmed/whole/coconut) etc.

- Low density lipoprotein, lecithin and phospholipids in egg yolk protect sperms by:
  - Forming a protective layer over spermatozoa
  - Replacement of lost/damaged membrane phospholipid
  - Reducing the phase change (liquid to gel) tendency
  - LDL especially phosphotidylcholine bind BSP (of seminal plasma) in diluents and prevent interaction of BSP with sperm (i.e. prevent capacitation and pre-acrosome reaction) and prevent membrane damage.
- In a certain time frame, BSP (bovine seminal plasma protein) binds to the sperm cell membrane and induces the capacitation in female genitalia. But during sperms storage, sperms are exposed to BSP for a longer duration and it causes loss of phospholipid and cholesterol which leads to damaged membrane (i.e. early capacitation like changes). LDL binds with BSP reversibly and protects spermatozoa from being capacitated during cryopreservation.
- Boar semen may be cooled up to 15^{0}C because poor lipid content in the sperm cell membrane makes it more sensitive to cold shock.

Cryoprotectant

- Cryoprotectants (CPA):
  - Minimize the intracellular ice formation
  - Reduces the solution effect
  - Improves cell survival during the freezing process
- Cryoprotective agents are of two types:
  a. **Penetrating** i.e. can cross the plasma membrane and acts intracellularly e.g. Glycerol, dimethylsulfoxide (DMSO), ethylene glycol (EG), propylene glycol (PG), dimethyl formamide (DMF)etc.
  b. **Non-penetrating** i.e. cannot cross plasma membrane and acts extracellularly e.g. polyethylene glycol (PEG), polyvinylpyrrolidone (PVP) disaccharides etc.
- During the freezing process, formation of ice crystals cause damage to the sperm cells. Cryoprotectants form hydrogen bonds with water which make water molecule less available i.e. less formation of ice crystals.

- Penetrating cryoprotectants bind with water molecules inside the cell (reduce dehydration rate) and thus prevent intracellular ice crystal formation while non-penetrating cryoprotectants increases the dehydration rate i.e. less water available inside the cell to form intracellular ice crystals during rapid cooling.
- The polar group of cryoprotectant molecules form H-bonding with polar head group of membrane lipid which stabilizes the sperm membrane during transition through critical temperature.
- Glycerol is considered as universal cryoprotectant and is used in cryopreservation of semen of various species. It acts:
    - Increase solvent viscosity inside the cell which reduce the solute crystallization
    - Salt buffering action
    - Increase glass forming tendency of the medium
    - Protect sperm cell protein from denaturation during freezing
    - Bind with water molecules and reduce freezing point (less ice crystal formation)
    - Formed ice crystals are of smaller size (prevent large sized ice crystal formation)
    - Reduce mechanical destruction due to ice crystals
    - Form channels/tunnels between ice crystal where spermatozoa lie without physical damage
- Permeating cryoprotectants like glycerol are more toxic to sperm than non-permeating cryoprotectants. Buffalo bulls are more affected than cattle bull spermatozoa.

Equilibration Of Diluted Semen

- It is the process in which sperm cell undergo adaptation with artificial diluter having egg yolk, glycerol and membrane stabilizer at low temperature (4^0C) for better survival during the freezing and thawing process.
- Equilibration is done at 4^0-5^0C for 4-5 hours by keeping diluted semen in a cold handling cabinet or equilibration chamber.

- During the reduction of temperature and equilibration, glycerol penetration into the sperm cells and the process of spermatozoa membrane stabilization occurs (i.e. protection from cold shock). It enhances the probability of better survival during freezing and thawing process with optimum PTM and fertilizability of sperm cell.

16

Semen Freezing and Thawing

Vikas Sachan

- After equilibration, a pre-freeze evaluation of semen is done. Pre-freeze motility should be more than 70% at 37^{0}C which makes the semen eligible for freezing.

Semen Packaging

- Semen is packaged in suitable container like pellets, glass ampoules, tupol (polyethene tubes) and semen straws.
- In the **Ampoule method**, semen is packed in glass ampoules and stored in LN_2. Here labeling and handling is easier with less chances of contamination. But more storage space, more time taking to thaw and the use of glass catheter for insemination make their the use less convenient.
- In **Pellet methods**, diluted semen is frozen by pouring it into the holes (0.1, 0.2 ml) made on dry ice (Solid CO_2 at 79^{0}c) and these pellets (frozen semen balls) are stored in LN_2. Here glycerolization is not needed as well as process is also quick but labeling is difficult and the chances to get contaminated are more. Handling pellets for insemination is also difficult (breakage or getting stuck to forceps during retrieval).
- In the **Tupol method**, polythene tubes are used for freezing semen at -196^{0}C. Semen is filled in polythene tubes with syringe and they are packed with electrical heating system in sizes of 3-4 inches each. Damage of sperm due to heat sealing, inaccurate volume, and spilling of semen at the time of cutting etc. are some disadvantages.
- Most recently **Semen straws** are used for the purpose of semen packaging because of the larger surface to volume ratio than pellet and ampoules. Semen straws are US/continental starw, German/minitube straw (packed with glass or steel balls) and French starw.
- French straws are large, medium and mini types. The most popular is French mini straw. Due to high surface to volume ratio of French

mini straw than medium, there is a rapid temperature change per unit time i.e. sensitivity of French mini straw toward temperature change is more. But due to efficient thawing and better recovery rate of post thaw spermatozoa French mini straws are widely accepted. Due to lesser diameter, there is the need of AI gun of lesser diameter which is easier to pass through the cervix at the time of insemination. More numbers of French mini straws can be stored in canisters due to thinner diameter.

- Sorensen discovered the semen straw in 1940. Cassou discovered French mini and medium semen straw in 1964. Danish starws were discovered by Sorensen (1940).

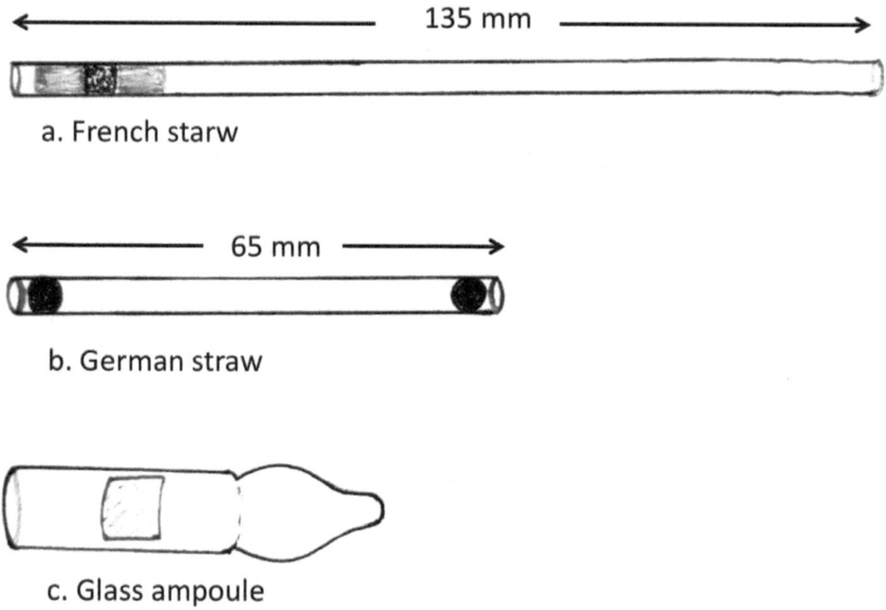

Semen Packaging
a. French semen straw b. German semen straw c. Semen glass ampoule

Straw Printing, Filling and Sealing

Semen straw filling and sealing is done at 4^0C preferably in the equilibration chamber.

- There are generally French mini straws are preferred for semen filling.

Sr.No.	Type of straw	Length	Diameter	Volume
1	French medium	135 mm	2.8 mm	0.5 ml
2	French mini	135 mm	2.0 mm	0.25 ml
3	German	65 mm	2.8 mm	0.25 ml

Identification Information On Semen Straw

- Semen straws are printed with straw printing machine with valuable identification information like
    - Bull identification no. (name or number & breed etc)
    - Semen station code
    - Batch no.
    - Date of production
    - Unique ID for every straw (if any)

Codes for Breeds

- Codes for some important breeds are-

S.No.	**Breed**	**Code**
1	Jersey	JY
2	Holstein	HF
3	HF cross	CB HF
4	Jersey cross	CB JY
5	Sahiwal	SAH
6	Red sindhi	RS
7	Kankrej	KANK
8	Gir	GIR
9	Tharparkar	THAR
10	Deoni	DEONI
11	Murrah buffalo	MBF
12	Surti buffalo	SBF
13	Bhadawari buffalo	BDBF
14	Nilli Ravi buffalo	NLRVB
15	Jaffrabadi buffalo	JBF

Straw Colour Codes

- Colour codes are also assigned for straws having semen of particular breeds.

S.No.	**Breed**	**Colour of straw**
1	HF	Pink
2	Indigenous cattle	Orange
3	Jersey	Yellow
4	HF CB	Green
5	Jersey CB	Salmon (Light orange to pink)
6	Sunandini	Blue
7	Buffalo	Grey

- If a straw of a particular colour is not available, a transparent straw should be used instead of any other colour assigned for another breed. Details should be mentioned in the records.

French Semen Straw

- French semen straw is made up of polyvinyl chloride; **PVC.** It has a **factory seal end** which contains **polyvinyl acohol powder** (PVA) while another other end is open. Semen straws are filled with semen manually (with the help of a suction pump) or with the help of automatic filling and sealing machine. As semen is filled into a straw from the open end and as PVA comes in contact with semen, PVA solidifies (gelatinous). Thus factory end is sealed. The open end of filled semen straw is sealed with PVA (manually) or by machine sealing and this is called as **lab seal end**.

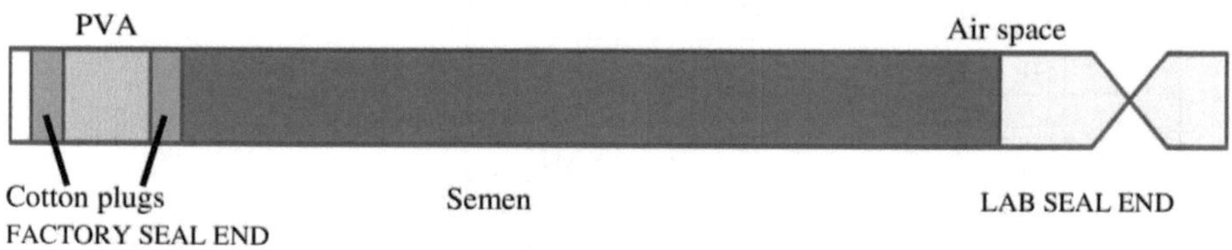

FRENCH MINI SEMEN STARW

- Air space is kept at the lab seal end for:
  - Non spillage of semen on cutting the straw at the time of AI
  - To adjust the expansion and compression of filled fluid during freezing and thawing
  - Probably to provide oxygen for sperm metabolism before freezing and after thawing

Semen Freezing

- After filling, sealing and equilibration (4^{0}C), semen straws are subjected to vapour freezing.
- During cooling and freezing, sperm cell undergoes **cryoinjuries**:
  - Phase change in membrane lipids along with Loss of cholesterol and phospholipids
  - Loss of calcium ions
  - Loss of ATPase activity of membrane

    - Capacitation like changes in sperm head (cryocapacitation) and pre acrosome reaction occurs
    - Increased solute concentration damages the sperm cells due to solvent movement
    - Ice crystals compress or damage the spermatozoa

- Sperm cells having a high **cholesterol to phospholipid ratio (C:P)** are more resistant to cold shock and cryoinjury e.g. rabbit, human etc. Sperm cells having low cholesterol to phospholipid ratio are more sensitive to cold shock and cryoinjury e.g. bovine, ram, buck, stallion and boar etc. C:P ratio in the spermatozoa plasma membrane is 0.83 and 0.88 in human and rabbit respectively. It is about 0.25, 0.36 and 0.43 in boar, stallion and ram spermatozoa respectively.
- Temperature from -5^{0}C to -35^{0}C is considered as a **critical temperature** for sperm cells. Sperm cells are subjected to maximum cryoinjuries as they pass through critical temperature (-5^{0}C to -35^{0}C) as well as recrystallization point of water (-130^{0}C) during freezing as well as during thawing.
- The **Recrystallization point** of water is the temperature below which water has no sufficient energy to form ice crystals.
- During freezing:
    - **a.** Water starts crystallizing (ice formation) in extracellular space
    - **b.** Solute concentration in unfrozen extracellular water increases i.e. osmotic pressure increases
    - **c.** Intracellular water undergoes exosmosis i.e. cell undergo dehydration
    - **d.** Intracellular ice crystals are formed
- Sperm cell may be damaged with fast as well as slow freezing:
    - **a. Slow freezing** – Large extracellular ice crystals area formed due to which more dehydration occurs (cell shrinkage). It leads to increased intracellular solute concentration (solution effect). Extracellular ice crystals damage sperm cells through compression.
    - **b. Fast freezing** – Less dehydration occurs which causes the quick formation of more numbers of small sized intracellular ice crystals. It leads to damage of sperm cell internal cytoskeleton.

- Fast freezing is more deleterious for sperm cell (more ice crystal formation) in comparison to slow freezing (solution effect). Cells which got damaged due to slow freezing can be revived up to a certain extent but cellular damages because of fast freezing cannot be reversed.
- The Freezing rate should be optimum (neither fast nor slow) i.e. combination of different rates of fall in temperature for different durations.

Methods of Freezing

- There are two methods of freezing :
  - Static freezing
  - Forced freezing

Static freezing

- It is also called manual freezing. It is conducted in an insulating wide mouth container (e.g. thermocol box).
- Semen straws are arranged on a rack and placed in the container having LN_2 (-196°C) in such a manner that straws are at the height of 4-5 cm (-180°C) above the level of LN_2.
- At this height, straws are exposed to LN_2 vapours and their temperature reaches to approximately -140°C within 10-15 minutes.
- These frozen straws are collected with cryogloved hands carefully and kept in pre-cooled goblets and canister. They are preserved by plunging in LN_2 in a cryocan.

Force freezing

- In this method semen is frozen with programmable biological freezer. There is a controlled rate of reduction of temperature (neither slow or fast freezing)
- The temperature of biological freezer is reduced to 4°C (equal to equilibrated semen straws) before placing straws into it for the freezing process with LN_2vapours. Further lowering of the temperature of semen straws in a programmable freezer is done in a controlled manner:

S.No.	Fall in temperature	Rate of fall	Time taken
1	4° C to -10°C	5°C per minute	2.8 minutes
2	-10°C to -100°C	40°C per minute	2.25 minutes
3	-100°C to -140°C	20°C per minute	2 minutes

Straw Storage

- These frozen straws are collected with cryogloved hands carefully and placed in goblets and canisters which are plunged into LN_2 filled cryocan.
- There are various mediums for cryopreserving the semen:

S.No.	Material	Temperature
1	Solid CO_2	-79^{0}C
2	Liquid air	-190^{0}C
3	Liquid nitrogen (LN_2)	-196^{0}C
4	Liquid He	-296^{0}C

- It was demonstrated that sperm survival at -196°C was virtually infinite, whereas the biological changes occurred at -79°C (solid CO_2) as well as solid CO_2 needs more frequent replacement. LN_2 is used extensively because it is easily available and economical.

Thawing of Semen Straw

- It is the process to bring the semen back from frozen state to liquid state by exposing it to a certain temperature for a defined duration.
- A Faster cooling rate used during semen freezing, requires a faster thawing rate for the liquefication of frozen semen.
- Rapid thawing is always better than slow thawing. In rapid thawing there is prevention of water recrystallization and less time for solution effect hence leser damages are there to sperm cells. Delayed thawing leads to solution effect, recrystallization of water and cold shock to the spermatozoa.
- Post thaw result is influenced by diluents, freezing rate, cryoprotectant, equilibration time and packaging system used etc.
- Various methods of thawing:
    - Shirt pocket thawing
    - Rolling between palm thawing
    - Air thawing
    - Ice water thawing
    - Warm water thawing
- Best post thaw motility is provided by warm water thawing. Various protocols for warm water thawing have been adapted:

  - 70^0 to72^0C for 5 seconds
  - 65^0C for 8 seconds
  - 45^0C to 50^0C for 15 seconds
  - **37^0C for 30-45 seconds (Practically applied and widely accepted)**
  - 4^0-5^0C for 5 to 30 minutes (ice water thawing)
- French mini/medium straws are thawed in **37^0C warm water for 30-45 seconds**. This protocol is widely accepted and practically easy for field inseminations.
- Thawing can also be done at 75^0C for 12 seconds. It provides better post thaw motility (PTM) but this thawing rate is not practical in field conditions.
- Change in "thawing temperature" has more effect on acrosomal integrity than thawing time. Overexposure to high temperature leads to the denaturation of sperm membrane proteins. Reduced "thawing time" leads to incomplete thawing i.e. post thaw sperm quality will be poor.
- Surface to volume ratio of straw influences the sensitivity of semen toward thawing temperature. Post thaw motility, acrosomal integrity and membrane integrity are found better in 0.25 ml straws than in 0.5 ml straws.
- Thawed semen should be used for insemination as early as possible. Holding temperature should be 37^0C and holding time should be <10-15 minutes. As much time lapsed, the chances of conception get reduced due to continuous deterioration of seminal quality after thawing.

17

Artificial Insemination (AI)

Vikas Sachan

- Artificial Insemination (AI) is the technique of deposition of semen (Fresh, Diluted or frozen thawed) in the female genital tract by artificial means for the purpose of conception.
- It is first generation biotechnology technique. It is the most economical and best technique which has been widely accepted for genetic improvement and production (calf crop and milk production) in farm animals.
- All the aspects like selection of bull, semen collection, semen evaluation, processing of semen, freezing, cryopreservation, thawing and insemination are equally important for making AI successful.

History Related To AI

- In the fourteenth century an Arabian placed a cotton ball in the mare's vagina and collect it after mating (soaked in semen) with a high-grade stallion; then by inserting this cotton ball into the vagina of desired mare and got it conceived.
- Anton Van Leeuwenhoek (Dutch scientist, 1678) and his assistant, Hamm, discovered sperm cell and called it as “Animalcule”.
- Lazzaro Spallanzani (Italy; 1780) was inventor of AI. He artificially inseminated a bitch and got 3 pups. He is known as “father of modern AI”. He concluded that sperm is responsible for conception. He also proposed later on that the life of spermatozoa prolongs on cooling (Cooling and freezing inactivate and warming activates the spermatozoa).
- Recto-vaginal method of AI was disscovered by Larsen (Danish Veterinarian) in 1938.
- 1st AI in cattle, mare and ewe was done by Ilya Ivanov (also written as Ivanoff) (1899) in Russia. Dr. Ishikawa was a pioneer of AI in Japan. The AI of swine was initiated by Ivanow in Russia in the early 1900s.

- 1st AI lab was established in Russia. 1st cooperative artificial breeding association was established in Denmark by Sorenson and Gylling-Holm in 1936.
- P. SampatKumaran started AI in India in 1939 at Palace dairy farm Mysore in Hellikar cows with semen of HF bulls. However, the first calf through AI was born in 1943 at the AAI (Allahabad Agriculture Institute) Allahabad, U.P.
- The classic rule referred to as the A.M. to P.M. and P.M. to A.M. system for insemination i.e. 1st insemination after 1st sign of estrous and 2nd insemination after 12 hours of 1st insemination, was established by Trimberger (1948).

Advantages of Artificial Insemination

- Artificial insemination is beneficial not only from the point of view of female animals but also for the inseminator, animal owner and managemental practices at the herd.
- From one ejaculate of an elite and proven sire (high genetic merit), hundreds of semen straws can be prepared i.e. more no. of inseminations can be performed in estrous females. While one bull can inseminate only one cow in natural mating.
- As more numbers of inseminations can be done in shorter time intervals, genetic improvement with the desired character can be brought very fast in a herd.
- More numbers of progenies can be obtained in shorter time intervals through artificial insemination i.e. progeny tested sires can be evaluated in the minimum possible duration.
- As bulls are reared in a scientific manner, disease conditions are diagnosed easily and quite earlier. Contagious diseases and other diseases are also checked for their transmission from male to female or vice versa because in AI, there is no direct contact of animals (Diseases may be transferred due to natural mating and animal to animal contact). The Spread of infections like IBR, BVD, brucellosis, JD, TB etc can be checked.
- There is very little chance for transmission of sexually transmitted diseases (Veneral diseases) because seminal examination is carried out at various steps during processing. Genetic diseases like BLAD (Bovine leukocyte adhesion deficiency syndrome, citrullinemia, DUMP

(Diuridine monophosphate deficiency syndrome), Factor XI deficiency etc can also be prevented from their transmission.

- Because of the great variation in size (very large or very small female), sometimes natural mating is not possible or very difficult. Artificial insemination overcomes this problem.
- Artificial insemination can be performed in animals in different geographical regions at different time. With natural mating it is very difficult.
- Artificial insemination makes the use of semen possible from a weak and physically unable proven sire (semen collection through electro-ejaculatory method) for inseminating females. In dog, AI is proven beneficial in cases of premature erection of the penis and poor coital lock etc.
- Artificial insemination also facilitates the use of semen from a sire that has died. Its semen (germplasm) is collected and preserved which can be used for insemination for a long period.
- Female animals, which are aggressive or non co-operative to be mounted by a bull, can be inseminated artificially.
- Cost may be high for transportation of bull from one place to another place for the purpose of introduction into a semen station or natural insemination. But its germplasm can be preserved and transported with minimum cost. It makes the use of the sire more efficient for insemination purpose with low transportation cost.
- Artificial insemination is the technique which makes the estrous synchronization protocols more successful for the purpose of better herd management.
- Artificial insemination ensures better safety for the inseminator as there is no bull to manage at the time of insemination.
- There is a reduced cost to rear the bull at the farm as lesser numbers of bulls can produce sufficient semen doses.
- AI is cheap and best technique with less work load.
- AI ensures better record keeping related to reproductive activities.
- Artificial insemination ensures the insemination with better quality of semen.
- AI allows the practical use of sex sorted (gender selected semen).

- AI may be proved efficient tool for the conservation of rare or endangered species.

Disadvantages of Artificial Insemination

- Artificial insemination has a lower conception rate than natural mating.
- There is a loss of genetic variation (genetic pool) from a population as semen from limited bulls is used for insemination.
- Genetic abnormalities, if any, in bull may spread very fast.
- Some males act as shredders i.e. they shed virus in semen without any clinical signs. These viruses transfer to female through insemination.
- Artificial insemination requires trained technical personnel and special equipments.
- Accurate estrous detection is required for evaluating the optimum time for artificial insemination.
- Strict hygienic and sterilized conditions are required to be maintained during AI.

Artificial Inseminatio in Bovines

- In bovines, Artificial insemination is done by vaginal method and recto-vaginal method. Vaginal insemination requires a large amount of semen and results in poor conception rate. On the other hand, recto-vaginal method is more commonly used. In the recto-vaginal method of insemination, AI gun is guided into the female genitalia by keeping a hand per rectally. Insemination is done in the uterine body.
- It is considered that AI in bovines should be done with AM-PM rule i.e. If an animal shows heat symptoms in the morning, it should be inseminated in the evening. If heat symptoms continue then a second insemination can be performed next morning. But now new concept is emerging that insemination should be done at the same time as estrous is observed and may be repeated after 12 hours.

Material Required

- There is a requirement of some materials like AI gun, AI sheath, semen straws in LN_2 cryocan, straw holding forceps (or tweezer), straw cutter, thawing tray or thawing machine, thermometer, tissue paper or cotton, full arm gloves, lubricant (e.g. paraffin), towel etc.

- The Purpose of AI gun is to contain the semen straw with a fixed AI sheath/covering (breeding sheath). AI sheath not only protects/ secures the semen straw in the AI gun but also checks the transfer of contamination between AI gun and female genitalia (Vice versa).
- AI gun is generally made of stainless steel and has a specific barrel size for specific types of straw. To fix the AI sheath (split or non split), The AI gun is of O-ring type, spiral type, protrusion (circular bump) type or universal lock type AI gun.
- If the thawing machine is not available, thawing tray should be used with warm water. Temperature can be checked with a thermometer.
- To cut the straw, It is recommended to use the straw cutter. Alternatively, sharp cutting scissors can be used.
- All AI equipments should be kept in a sterilized AI cabinet/chamber. Regular cleaning and sterilization is recommended.

Steps Involved in AI in Bovines

The Inseminator should have to take the history about the female reported in estrous related to its last parturition, any post-partum complication, inter estrous interval, heat duration, heat symptoms and heat intensity etc. The animal should be restrained and rectum should be emptied with lubricated gloved hand. Its reproductive organs should be palpated rectally to rule out any sign of pregnancy (may be gestational heat) & to confirm the estrous.

- Tie or hold the tail aside.
- Clean the vulva and perineal region with water and make it dry.
- Keep the AI gun ready with plunger out (about 1/4th) along with the AI sheat
- Open the lid of cryocan and put it aside vertically inverted.
- Lift the canister having the desired semen straw upto the level of the neck or frost line of the cryocan (the straw should be hardly visible).
- Take out the semen straw, wave it in the air gently or wipe it with tissue paper and put it into thawing water immediately. It should be at 37^{0}C for 30 – 40 seconds.
- After thawing, the Semen straw should be held upright with forceps and it should be wiped to remove any trace of water over it.

- Load the semen straw in the AI gun by placing its factory seal end inside the AI gun.
- Strike gently over the lab seal end to make sure that all the air inside the straw should be at this end.
- Cut the straw with a straw cutter.
- Place AI sheath over the straw holding AI gun.
- Put left hand in the rectum of the animal holding the cervix and hold the AI gun in right hand.
- Make the vulval lips apart with the help of an assistant or by pressing the rectal opening downward.
- Direct the AI gun to the uterine body and ensure its correct position. Semen is deposited by pushing the plunger.
- Remove out the AI gun back slowly and dispose the AI sheath with hand gloves properly.

Precautions During After Artificial Insemination

- Any stress to the animals during approach to the hospital (running, jumping etc), restraining (beating), rectal palpation (sudden and forceful) etc may lead to the rise of stress hormone (adrenaline/cortisol) which suppresses ovulation and leads to poor conception. Animal should be given rest of about 15-30 minutes before and after insemination.
- The Packet of AI sheath should be torn only from the end opposite to the one which is to be inserted into female vagina. "Single sheath within single seal" is better to use.
- The Canister should not be lifted above the start of neck or frost line of the cryocan to avoid thermal shock to the spermatozoa.
- Tweezer or straw forceps should be cooled before picking up the semen straw from the cryocan by exposing it to the LN_2vapours within the cryocan.
- If the straw cannot be picked within 10 seconds of lifting the canister, it should be immersed back in LN_2 and it should be lifted again after 5-10 seconds to retrieve the semen straw.
- Semen straw, top **3/4th** portion of AI gun or AI sheath should not be touched with barren hand to avoid any transfer of contamination.

- Straw should be cut with a straw cutter by keeping it in upright position. Scissors should be avoided, as it may crush the cut end of the straw i.e. no sharp and neat cut. If scissors is to be used, straw should be cut by keeping it at the level of eye. Cut should be even and horizontal otherwise during insemination; semen will leak within the AI gun.
- Thawing should be done by keeping the straw horizontal to ensure thawing of the whole straw at same temperature. In the case of thawing machine, whole the water is at same temperature, so semen straw can be kept in any plane.
- Insemination should be done as soon as possible (maximum within 10-15 minutes) after thawing of straw.
- Loaded AI gun or AI gun after thawing is not recommended to be in contact with any contaminated surfaces.
- Ensure the positioning of the AI gun at right place before semen deposition. To ensure the position of gun just before semen deposition, the inseminator should palpate the tip of the AI gun and barrel/body of AI the gun for a few inches.

S.No.	Possible position of AI gun	Tip of the AI gun (Palpable or not)	Body of the AI gun (Palpable or not)	Insemination
1	In vaginal cavity	✓	✓	Not correct
2	At Fornix area	✓	✓	Not correct
3	In the cervix	X	✓	Correct
4	In uterine body	✓	X	Correct
5	In Urinary bladder	✓	✓	Not correct

- Semen should be deposited slowly and with uniform speed in the uterus. It should not be deposited too farther in the uterus to avoid unicornual semen deposition.
- If efforts do not lead to the right positioning of AI gun in the genitalia, semen may be deposited at external Os or in mid cervix. More manipulation by striking the AI gun for a longer duration may lead to unwanted injuries within genitalia.
- After insemination, tip of the AI gun should be examined for any blood or pus (infection) traces.
- Clitorial massage for 8-10 seconds and pouring water on the hind quarter after AI improves the conception rate. It improves the contraction and relaxation of genitalia which support sperm transport and so improves fertilization. Administration of oxytocin also is found to improve the conception rate but it is not recommended.

- AI gun and straw cutter or scissors should be cleaned and sterilized after every insemination to avoid the transfer of any possible infection from one to another animal. Whole of equipments and AI kit should be sterilized once a week.
- Records related to insemination along with straw information should be recorded properly by writing or by suitable digital application.

Artificial Insemination in Doe and Ewe

- Cervical and uterine insemination is preferred with chilled or frozen semen. Intrauterine insemination can also be done with intra-cervical or laparoscopic technique.
- Doe or sheep (which is in heat) is lifted with its hind quarter in an inclined position by securing its hock joint. The vaginal opening should be faced toward a light source. A sterilized tubular speculum (may be of glass) is inserted into the vagina till cervical opening. Inseminating catheter or AI gun is inserted through this speculum and directed into the uterus through the cervix to deposit the semen.
- Pregnancy rates are higher when semen is deposited directly into the uterus through laparoscopic technique. In this technique, semen of amount 0.05 ml is sufficient for insemination (25-40 million spermatozoa). However, this technique is costly and requires skilled personnel.

Artificial Insemination in Mare

- In mare, vaginal insemination is preferred because cervix is not so muscular to palpate per-rectally. Also rectal muscles are more fragile to manipulate.
- Precaution should be taken that perineal area should be cleaned and dried. The tail head should be wropped with bandage and kept aside at the time of insemination (hairs of tail are quite tough and may cause irritation and inflammation if trapped in the vagina or rectum).
- Semen is deposited in the uterus with the help of a long catheter using gloved and lubricated hand through the cervix with the guidance of index finger.

Artificial Insemination in Sow

- The Vulvar area should be cleaned and dried. Environment should be kept calm and stress free for the sow.

- Artificial insemination in sow is done with a spirette (catheter with spiral tip) or foamed catheter (catheter with foam at the tip) attached to a semen bottle.
- Catheter should be passed through vagina and locked at cervix with rotating action. Then semen bottle is attached to it and little squeezing is done. Semen is sucked into the uterus because of uterine contractions stimulated due to cervical pressure.
- The Catheter should be taken out (by rotating) gently to avoid backflow of the semen. Female should be kept calm and stress free for about half on hour after the insemination.

Artificial Insemination in Bitch

- Frightens or aggressiveness of either the male or female, inability to copulate or improper locking etc. are some factors on which artificial insemination is preferred.
- Semen should be inseminated immediately after collection of ejaculate. In delayed insemination, it should be preserved in extender containing cooling protectant like egg yolk, and can be transported at the temperature of 5^{0}C.
- As the life span of frozen-thawed semen is very short, intravaginal insemination is preferred with chilled or fresh semen.
- 2-3 inseminations are required at the interval of 48 hrs starting from 48 hrs of cytological estrous till the end of this phase.
- It is always better to do AI according to plasma progesterone level. Following regimen can be followed:

S.No.	Plasma progesterone concentration	1st AI (days from progesterone analysis)	2nd AI (days after 1st AI)
1.	5-8 ng/ml	2 days	1 days
2.	9-12 ng/ml	0 day	2 days
3.	12-20 ng/ml	0 days	1 day
4.	>20 ng/ml	0 day	-

- Semen is deposited at the anterior vagina (with a catheter) or in the uterus. Generally, semen is deposited into the vagina because it is difficult to pass on the catheter into the uterus due to "Z" shaped cervical lumen.
- With fresh semen at room temperature, up to 3 ml, 5 ml and 8 ml is enough to intravaginal inseminate in bitches of smaller breed, medium

breed and larger breed respectively. Intrauterine insemination can be done with 1 ml, 2 ml and 3-4 ml in the case of small, medium and larger breed.

- After depositing the semen with the help of a catheter into the deeper area (anterior vagina) of vagina, "feathering" is recommended. **Feathering** is the process to give stimulation into the dorsal wall of vagina by doing gentle backward motion. It stimulates the secretion of oxytocin due to which uterine contractions improve to facilitate the spermatozoa traverse through female genitalia.
- "Osiris gun" can also be used for intravaginal insemination in bitches. This gun is made of a flexible tube having inflatable balloon at anterior portion which mimics, when in inflated condition, like erected penile bulb of the dog as well as prevents backflow of semen.
- Intracervical or intrauterine insemination in dog is done with non surgically "**Norwegian method**" (endoscope-assisted intrauterine technique) and "**Newzealand endoscopic method**". Intrauterine insemination can also be done with a surgical method (laparoscopic) but it can be considered unethical.

S.No.	Animal	Time of Insemination
1.	Cattle	0-12 hours after onset of estrous (Repeat after 12 hours)
2.	Buffalo	0-12 hours after onset of estrous (Repeat after 12 hours)
3.	Mare	Start at 2nd day of estrous then alternate days
4.	Sow	18 hr after onset of heat (repeat after 12-24 hrs)
5.	Ewe	Mid to late heat (12-18 hr after onset of estrous)
6.	Doe	12 hr after onset of estrous
7.	Bitch	From 2nd day of onset of estrous (2-3 inseminations at interval of two days)

S.No.	Species	Site of insemination	sperm cells (Volume of diluted semen)	sperm cells (Volume of frozen semen)
1.	Cow or Buffalo	Internal os or uterine body	10 million (1 ml)	20 million (0.25 ml)
2.	Mare	Uterine body	500 million (10-20 ml)	800-1600 million (0.5-5 ml)
3.	Sow	Cervix or uterus	50-100 million (100 ml)	2500-4000 million (80-100 ml)
4.	Ewe/Doe	Uterus	100 million spermatozoa (0.05-0.1 ml)	100 million (0.25 to 0.5 ml)
5.	Dog	Anterior vagina	200 million (5-10 ml)	250 to 700 million (2.5 to 3.5 ml)

18

Surgical Procedure on the Male Reproductive Tract

Vikas Sachan

- There are some methods to control male reproductive behaviour for better breeding and managemental practices i.e. Vasectomy and Caudal epididymectomy, Penile surgeries, and castration etc.
- Surgical and non surgical interventions in the male reproductive system are carried out for:
    - To make teaser bull
    - To make bull sterile
    - To prevent the transmission of venereal diseases
    - Population control or Breeding management
    - To rectify any ailment or deformity

Teaser Bull

- A male animal whose capability of spermatogenesis as well as libido is maintained but it is not able to impregnate a female.
- For the preparation of teaser bull or Gomer bull, Male animal should be selected well before the start of breeding season. Bull should be young vigorous and fertile. It should be of moderate adult size with mild temperament. Heavy and large sized bull is tough to handle, needs more time for surgery, has chances of more hemorrhage, and may have more post surgical complications.

Vasectomy

- In vasectomy, only the passage of sperm transport is blocked. Male behaviour of the male e.g. detection of estrous female etc is intact due to maintained level of testosterone.

- General line of procedure-
  1. The animal should be off-feed for 24 hours
  2. Restraining of animal should be on lateral or dorsal recumbency
  3. Sedation with inj xylazine (@ 0.05-0.1 mg/kg bw IM or IV)
  4. The Site should be prepared aseptically and local anaesthesia should be used.
  5. 2-3 cm incision is sufficient at the neck of scrotum at median raphae should be taken. Two separate incisions may be taken on the neck of testicle at palpable spermatic cord.
  6. 2-3 mm thick vas deferens is identified and separated from other spermatic cord structures. It is elevated at the level of incision site and two ligations 3-5 cm apart are put. The portion of vas deferens is cut-off taking slight margins from ligations.
  7. Incisions may be sutured properly or sometime it may be left for open healing (if the incision is quite small)
  8. Antibiosis and anti-inflammatory should be administered for the next 4-5 days with daily dressing.
- The Vasectomized male animal should not be used for atleast next 30 days as spermatozoa may be there at distal portion of the genitalia.

Caudal epididymectomy

- General line of procedure
  1. Site preparation and other measures are the same as vasectomy
  2. The testicle should be squeezed on the ventral side and cauda epididymis at the bottom of testicle is palpated.
  3. The tail of epididymis is exposed through a small incision of 2-3 cm (under local anaesthesia).
  4. Single ligation can be put or a part of cauda epididymis is separated out in the same manner as of vas deferens in vasectomy.
  5. Wound management and post operative care should be carried out aseptically as routine procedures.
- With vasectomy and caudal epididymectomy, penile intomission is not checked that's why there is no prevention of transmission of venereal diseases.

Penile prepuce translocation (PPT)

- It is the formation of a new tunnel subcutaneously displacing the penis and prepuce from natural opening to an alternate exit at the lower left / right flank region. It does not prevent the protrusion of penis but intromission is checked due to deviated opening (No alignment against the vaginal opening).

Penopexy

- It is the process to fix the penis by creating adhesions between the tunica albuginea of the penis and ventral body wall (linea alba). It prevents protrusion as well intromission of the penis.
- A part of the penis is exposed surgically 2-3 cm away from a line joining preputial opening and scrotal base. Scarification at dorsal surface of penis (tunica albuginea) and linea alba is done. Both the scarified tissues are then sutured to create adhesion in order to fix the penis. The surgical incision site (Subcutaneous muscles and skin) is closed with standard protocol. Post operative care is done with antibiotics, anti-inflammatories and daily dressing aseptically.
- Penile fixation by adhesion may satisfactorily occur in 5-6 weeks.

Penectomy

- It is the surgical removal of penis.
- Vasectomy, Caudal epididymectomy, PPT, Penopexy, and Penectomy do not affect the libido of a bull as testes are intact. These bulls are called teaser or gomer bulls.

Castration

- It is the process of removal or destruction of the testes by physical, chemical or surgical methods. It is also called as orchidectomy or orchiectomy.
- In the animals whose testes are removed (castrated male), Spermatogenesis as well as testosterone is not produced. It makes the animal docile and easy to handle.
- Calves should be castrated before 8 weeks of age. More the younger calf, less stress will be there. Castration in older calves requires more care, management and post-operative care.

Physical methods

- Elastrator band – It is a tool to apply on elastic band over the neck of the scrotum in younger calves of less than 3 weeks of age. Blood supply is checked and the scrotum along with testicles are shed off from the body in due course of time due to necrosis.
- EZE or Callicrates – These are tools to apply latex bands in larger calves.
- Burdizzo castrator and emasculator – These are applied to crush the spermatic cord.

Chemical method

- Intertesticular injections of NaCl, lactic acid, a combination of $CaCl_2$ + dimethyl sulphoxide or hydroxyl propionic acid etc can be used for castration.
- Immunocastration
- Hormones or antigens to suppress the GnRH secretion are used as immunocastrating agents. But the results are controversial.

Surgical method

- It is the removal of testes with open surgery under local or general anaesthesia depending on species. Blood vessels in the spermatic cord are ligated and testicles are removed by cutting the spermatic cord with standard surgical procedure.
- Incisions on the neck of scrotum are generally closed left with little opening to drain out the inflammatory fluid. But sometimes in cases of necrosis or degeneration, scrotal sacs were also removed.
- Castration in equine is generally called **"gelding"**. Incisions are kept generally open for open wound healing in equines.